KU-377-183

CAPTIVE WILD

BOOKS BY LOIS CRISLER

CAPTIVE WILD
ARCTIC WILD

CAPTIVE WILD

by Lois Crisler

W. H. ALLEN · LONDON · 1969

First British edition, 1969
Printed in Great Britain by
Fletcher & Son Ltd, Norwich
for the publishers
W. H. Allen & Co, Ltd,
Essex Street, London WC2.
Bound by Richard Clay
(The Chaucer Press) Ltd,
Bungay, Suffolk

491 00402 8

Contents

CAPTIVE WILD

1 *Travel*

"I WISH you could have seen the wolves," I said to friends seeing us off at the airport at Fairbanks, Alaska. We had five tundra wolves, nearly full grown, that we had acquired during the course of a long photographic mission in arctic Alaska. The mission was over now and we were en route with the wolves to our home in Colorado. Permission to take the wolves into that state had just arrived. Until it came we had not known where we would take them and had thought we might have to give up Crag cabin, our Colorado home, in order to go where we could keep and care for them.

We went out to the plane for Seattle, our first stop, and boarded it to find the wolves in the passenger compartment and the stewardess taking people up front, one at a time, to look at them. The front seats had been removed to make room for their travel boxes and that of our dog Tootch. The animals had been placed in the cabin because the baggage compartment was not pressurized and the plane would fly at a high altitude.

The wolves were terrified. They lay motionless at the back of their boxes, their eyes brilliant with fear. They were making themselves as inconspicuous as possible. Reared at our lonely camp in the Arctic, they had known, until this journey began, only the great tundra and ourselves, my husband Cris and me. Fortunately their travel boxes gave them a measure of concealment: they were of plywood, with a screen panel at the front end only. (A floor of wire netting above the wooden one kept fur from being soiled by droppings.) The wolves must have suffered from the warmth of the cabin, for the month was October and they already had their full winter coats for the Arctic—deep soft blankets of wool underlying their guard hair. But they did not move nor utter a sound.

The dog Tootch behaved differently. She was a vivid red-gold sled dog, an ex-lead dog of Eskimos. She crowded the front of her box, her friendly dark eyes full of distress and appeal. After take-off she sent her sled-dog wail soaring above the noise of the plane. The stewardess hastened to bring her water, which the dog drank. It did not occur to me that the girl had any other motive for her concern to quiet Tootch than a right-minded love of animals. Tootch yowled again and the girl brought her food. Tootch ate.

The wolves must have been thirstier than she but they did not move to touch the water the stewardess gave me for them. They lay like shadows, except for their watchful, fear-bright eyes. All alike were practicing the wolf technique of silent passivity in the face of overwhelming odds. A stranger could not have told one from another, but to us each wolf was a distinct personality, differing from its fellows in physique, bearing and disposition.

The biggest and rangiest of the wolves, Mr. Arctic, was a lordly animal. His brother Barrow, less weighty in character, was stocky and effusive. Miss Tundra was reserved toward human beings, giving her devotion to the dog Tootch. Her sister Killik was a big, plain, "ordinary" wolf, so I thought.

The gayest, prettiest and most winsome of the wolves was Miss Alatna. She was slender and graceful in build. Romping, tireless, devising games, she usually led the pack. All the wolves were friendly but she was the most human-oriented.

Confiding toward us, she was confident of life. It seemed more unhappy for Alatna to withdraw into fear than it did for the others.

It was only natural that we should know these wolves as individuals. We had reared them from baby puppies, only three weeks or so old. They were five months old now, and during those months they had become part of our family, part of our lives. Much of the time they had been absolutely free. Running collarless and free, they shared hours-long daily walks with us over the tundra. The proud look of freedom showed in their eyes and stance. Would the terror of this journey efface it?

I could not comfort them. I knelt in front of their travel boxes and spoke gently but they gazed at me with brilliant, unchanged eyes. I returned to my seat. As we flew southward with our precious, unwieldy cargo, full of living misery, I thought about the first wolf I had ever longed to know. She had seemed unutterably mysterious and unknowable.

We came upon her in the course of our work. We had been sent to arctic Alaska by a Hollywood producer, nearly a year and a half before, to photograph the North. On a cool gray day in our first arctic June we were backpacking north of the Brooks Range. We stopped at the sight of an animal crossing a marsh at our left. A wolf. The wolf caught our scent, paused and looked, but not quite at us, then trotted onward and up onto a low ridge at the side of the marsh and stood there against silvery clouds. Her tail hung straight down. To me she looked lonely. But no human, I thought, could know.

She sat down for a while, like a slant rock on our low skyline. But presently she rose and trotted back along the top of the ridge. Then we noticed the wolverine. He was angling up the side of the ridge as if to go back along its top.

Acting concerned but quiet, the wolf trotted to and fro in front of him until she had headed him off and started him downward the way he had come. But promptly he turned and climbed upward again. Again the wolf stopped him—temperately—and headed him back downhill. Were her pups back along the ridge? The wolverine returned. Both animals were bent on having their own way.

Once more the wolf pointed the wolverine downhill and this time he kept going; he lolloped downward and away. The wolf went back to her lookout, as to an accustomed place, and sat down.

Standing beside willows below, oblivious of my pack, I felt a quiet deep longing to know her, not only information-about but awareness-of, as one would know another being. No words, just this quiet inward turning of myself toward her. But the wolf so near up there looked remote as a being on another planet, orbiting near mine for an instant, to whirl irretrievably away into space.

Only two weeks later we received our first wolves, Trigger and Lady, puppies taken from their wild den by Eskimos at the cost of their siblings' lives. So evil and good were mingled in our enterprise from the start.

We reared Trigger and Lady in close intimacy with us. When they were grown we set them free on the tundra, to hunt for themselves, make acquaintance with wild wolves and return to us for support and affection if they wished. They did return almost daily: we were their family; our camp was their home. It was also that locus indispensable to far-ranging wilderness animals few in number but social—their rendezvous. Here if separated, as rarely happened, they could await one another.

Our camp was a "crackerbox" of plywood and plastic glass, on the brink of a mesa near the junction of the Killik River and a tributary. Looking for miles up both mountain-lined river valleys, it commanded vast sweeps of sunlit tundra and mountainsides.

A year after Trigger and Lady entered our lives we acquired the five wolves, a full litter, that we were now flying homeward with us. Like our first wolves, they had been taken as puppies from a wild ravaged den. To our delight Trigger and Lady instantly adopted them and undertook to support them. So for a while the new puppies enjoyed a near-normal life, cared for by a male and a female wolf. When Trigger and Lady, by chances of the wild, disappeared and we alone were left to act as foster parents to the five, they remained docile and affectionate toward us as always, returning cheerfully to camp from their long free hours on the tundra and sometimes preceding us home.

Now we were taking them by a long hard way toward our real home, Crag cabin, in Colorado.

It was evening when we arrived in Seattle. Friends met us and accompanied us to the railway freight shed to which the travel boxes were taken. From Seattle to Denver we were to go by train because the animals could not be taken on a plane with us but would have been routed by way of San Francisco, possibly without water and certainly without the sanity-preserving touch of something familiar.

In the cavernous, dimly lighted freight shed Mr. Arctic was taken from his box on a chain and displayed to the group of our friends. His tail was clamped to his belly, his eyes were brilliant and hopeless, he was passive and silent. After he was restored to the poor shelter of his box, one friend, Carol, detached herself from the group and kneeling in front of his box spoke directly to him in a low voice. I knew she could not allay his terror but I felt grateful for her compassion.

On the train to Denver we kept Tootch in our compartment and at every long stop took her on leash to the baggage car and led her in front of the travel boxes of the silent wolves. She had a peculiar relationship with the wolves: she was hostile and cocky toward them, yet they courted her sweetly as the elder animal. She had been jealous of them as puppies, later somewhat afraid of them as they grew bigger than she; and at all times she had sought to maintain her lead-dog supremacy over them. She bit their muzzles if the wolves offered to kiss her. So the wolves courted her cautiously and briefly. But Tundra, the puzzling, odd-ball wolf, courted her with reckless persistence; more than once as a puppy she had lain alone all day, whimpering softly to herself now and then from the pain of a swollen, bitten muzzle. Afterward she had courted Tootch as sweetly as ever. I wondered whether she missed her mother more than the other wolves did. Now she and the other motionless wolves in their travel boxes could at least see and catch the scent of cocky, edgy Tootch, an animal important to them.

Back in our compartment after these visits to the wolves I thought with dread of the hard time awaiting them even after the hardship of this traveling was over—the time while a pen was being built for them at Crag cabin and they would have to be chained to stakes.

The reason why we were subjecting them to all these miseries was not that we wanted them as pets nor as objects for study, but simply that death for them would have been the only alternative to bringing them along home with us. We could not do what we would most have liked to do—turn them loose on the tundra to live free. They would have starved. They were big but still puppies at heart. Wolf puppies run with both parents until nearly a year old, receiving support and experience in hunting. We ourselves had supported Trigger and Lady during their transition to living wild. Moreover, it had been spring when we set them free. Now the arctic winter was approaching. At the crackerbox in the Killik the caribou herds had long since passed, going southeastward across the Brooks Range for the winter. Even an experienced wolf could hardly have wintered over here. It was not possible for us to abandon our young wolves to slow death by starvation.

We could have given them quick death by means of sleeping pills followed by a bullet. Perhaps that would have been the more merciful course. It would not have been irresponsible. But in a last-minute decision we had assumed a more long-lasting responsibility for the wolves rather than take their lives.

On the evening of our arrival in Denver we were met by a friend who had brought our four-wheel-drive pickup to the station. Into it were loaded the travel boxes, Tootch and our gear, and we set out on the last leg of our journey from the Arctic, the long drive toward Crag cabin, in the Tarryall Mountains southwest of Denver.

It was two A.M. when I unlocked the long-unused front door of our little log cabin. In the beam of my flashlight the polished tan linoleum floor looked as big as a ballroom floor. This feeling of its size was a hangover from living for the past months in tiny quarters—a tent or the crackerbox.

For this one last night the wolves had to remain in their travel boxes: we could not stake them out until daylight. We set the boxes on the ground and went to bed. In the pure dawn twilight the wolves, emboldened by the profound stillness, began trying to escape. The noise wakened us. We rose and started the process of staking them out. The hard time I dreaded was beginning.

2 *The "Hard Time"*

CRAG CABIN stands in a small wooded draw at the foot of a southwest flank of the Tarryall Mountains. These are low mountains compared with the many 14,000 footers in Colorado; they rise only to 11,000 and 12,000 feet. They are midway between the main range of the Rockies and the abrupt plunge of the cordillera downward to the high plains to the east. Down there along the foot of the mountains are strung the cities of the plains—Denver, Colorado Springs and others—each under its smother of tan air. That sullied air and the dust of the plains rarely climb through passes and over ridges to the lonely little draw where Crag cabin stands.

The draw lies at the exact margin between wild and tame; it belongs to the wild. Around it on three sides rise the mountains or great shoulders from them. The fourth side is formed by a low wooded ridge that flows off the mountainside and turns to enfold the draw, cutting it apart from the weary overgrazed cattle pastures out along the Tarryall valley.

Leaving the pastures, a one-track road rounds below the point of the ridge and enters the draw by way of an aspen flat. At the left of the flat it fords a brook in a "gate" of thick dark firs and winds steeply upward, curving along the inner side of the enfolding ridge. Where it rises above the tops of the trees—aspens, firs, spruces—that grow from the floor of the draw you can look across at a bench on the opposite side. Rising from this bench, the mountainside leans back in giant stairs of naked granite and patches of forest, going upward to the skyline crags two thousand feet above.

The road ends on the inner corner of a low shelf niched in the bend of the ridge and there, when first we saw the draw, stood a crib of logs, weathered black and rough, laid up decades before by some forgotten prospector toward a cabin; bits of purple fluorspar sprinkled the ridge side above the shelf, and there was an old prospect hole up there. This was an abandoned mining claim. The log crib had never been finished. Open to the sky, it was carpeted with wild grasses and gray, pungently sweet sage herbs.

Crag cabin was built inside the oblong of the old log crib: one room of knotty pine, with a great oversized front-and-only door, topped by six mountain-facing panes. On the down-draw side of the cabin a big picture window was let into the old logs. It overlooked the near end of the floor of the draw and part of the rugged hillside leading upward to the bench; it framed, beyond them, the wooded craggy triangle of a mountain that seemed to barricade the foot of the draw but in reality rose beyond the aspen flat and a ridge on its far side.

On the morning of our return home, as we worked, still in the mountain shadow, to stake out the wolves, it felt strangely comforting to be held here in the hollow of the mountains. I had felt no strain at the vastness of the arctic tundra, but strain there must have been for here now was this feeling of comfort at being enclosed.

Stakes were driven on the hard bare plaza, as we called that part of the shelf lying in front of the cabin. There was room here for only three wolves, as they had to be far enough apart so that their chains would not tangle. The other wolves were fastened on the side of the ridge at the left of the plaza, Miss Tundra to the big red trunk of a ponderosa pine.

This first day at Crag cabin was frightening for the wolves, for the rumor had spread that we were bringing home wolves and visitors came by jeep or on horseback to look at them. The wolves were exposed helplessly without even the shelter of their travel boxes. When the last visitor of the day arrived, a woman on horseback, Tundra and Alatna broke loose in desperation and disappeared. I asked the woman to leave, for I knew I could not get the wolves back while she was here. After horse and rider had disappeared down the road, I started a howl, in which the chained wolves gladly joined. A howl was like a community sing; it was an outlet for their feelings. Alatna, who dearly loved a howl, answered from the mountainside and came bounding down out of the woods to join in. Presently Tundra came too. Within ten minutes I had them both tied up again.

That evening we decided that to keep the wolves from going crazy with despair during the "hard time" while their pen was being built we would simply turn them loose, one at a time, each wearing a choke chain and dragging a six-foot chain to facilitate recapture. We supposed all we had to fear was a drag chain snagging between boulders and holding the wolf a prisoner, silent and unfindable, perhaps miles from home. I had forgotten one thing—an old feud, begun on the tundra weeks before between odd-ball Tundra and her sisters. It had smoldered and then, apparently, died.

For a day or two all went well. The released wolf played with its fellows or ran off exploring but soon returned, just as we had anticipated. The pull of the wolves to stay together was very strong. Then one morning as I released Tundra from the pine tree she gave me the slip without her drag chain and flew straight down at Killik, staked on the plaza. I jumped into the fight barehanded, trying to haul Tundra off. Not until the backs of my hands were bitten—inadvertently—did I remember the spare choke chain and leash. I ran and brought them. In the whirling melee of fangs and fur I got the choke chain, I thought, over Tundra's head and pulled with all my might. Clear grave wolf eyes searched mine—Killik's eyes. It was around her neck that I had slipped the spare choke chain; I was strangling her between it and her stake choke.

Killik apparently thought I had jumped on her, siding with Tundra. She was timid of me after this. Adding to her timidity was a fact I did not know yet: Killik already distrusted me. Not until long afterward, when I chanced to go through the stills I had taken of the wolves on the arctic tundra, did I gradually perceive that in every group picture taken after the middle of August, while the other wolves went about their affairs, one level-eyed gray face, Killik's, would be steadfastly regarding the camera, that is, me. What had I done to make her suspicious of me? I could never know. Would she have attacked me? No! Far more likely she feared I might attack her and so was on her guard.

One day the howling of the staked wolves called me outdoors to see Killik standing chain-caught between boulders up on the sunny mountainside. Climbing toward her I thought despairingly, If only it had been any other wolf. I did not know whether, when in trouble, she would even let me approach her. But she did. She stood still while I loosened the chain.

I rotated the wolves, kept house and prepared meals for human beings and the six big animals. It kept me busy. The best I could manage for the wolves was a free hour apiece, twice a day. During each free hour I kept track of the whereabouts of the free wolf if it stayed around, and near the close of the hour I watched for a chance to lay my hand unobtrusively on the end of the drag chain. Each free wolf except shy Killik entered the cabin at will; the door usually stood open.

Tootch, who was not tied up, did not fraternize with the wolves, but the free wolf, especially Tundra, often approached to court her and was rebuffed as always.

Quickly the wolves became accustomed to the routine of free hours. Seeing Barrow taking off one morning near the end of his free hour, they howled. This alerted me in time to call Meat! before he had gone out of hearing. Back he came, up the road.

Broad-backed as a circus horse, he entered the door, tilting his head, his eyes beaming in a lovesick way. Barrow was a luscious cuddle wolf. His deep-furred jowls had the look of Lord Dundreary whiskers because of a black streak of fur slanting down from the outer corners of his eyes. So big, with no inhibitions about tables—nothing sacred—he made the cabin feel crowded and paper-frail. He could never get it out of his furry head that once in the Killik he had got a whole baking, two loaves, of fresh bread from our crackerbox there. I fulfilled my promise and then restaked him.

Each wolf but Killik had some little distinguishing mark. Alatna's was a whorl in the soft fur of her forehead. Tundra's was gray "spectacles." Arctic had a rough nose tip; Tootch had bitten it off when he was a puppy and it had healed wrinkled. But we did not need marks to tell the wolves apart; long since we had learned to distinguish them afar by their build and bearing.

Each wolf awaited me in his or her own particular way as I went to release it. Killik was the tail-wagger, but not for me! —for the free wolf when she saw it approaching. Alatna was the "patty paws." Leaning against her chain she made quick dancing pats with her front paws, as she had patted them alternately when a puppy, nursing another wolf's ear, or waiting happily among her fellows for the gate to open for the long free daily walk. She was a singing, dancing wolf. She threw herself into a howl more ardently than the others.

Barrow dug with both front paws together, impatiently watching my approach. Tundra was reserved toward people, but she stood on her hind legs and flung an eager paw out to the side as I approached. Arctic nearly broke my nose, jumping eagerly to kiss my face in his excitement over release at hand. Free, he could not run instantly but had to pause a second to exchange felicitations with me.

Similarly in the old days Trigger and Lady when chained had awaited us differently. Black-eyed and crying, Lady would throw herself against her chain, while Trigger had sat graceful and silent, waiting.

Arctic reminded me in so many ways of Trigger that sometimes I absently called him Trigger. He was lordly like Trigger. We had him staked near the door, at one end of the low plank deck along the front of the cabin. He required to examine everything we carried in or out the door; we held it for him to do so, even to an armload of wood for our little cooking and heating stove.

Together we had gone to release him for his very first free hour. After he was released he had stood still beside his stake. We had walked away, coaxing. He followed, but only to his familiar chain length—he wasn't going to be jerked! He went back to his stake, touched it with his nose, turned his head and looked at the chain end lying on the ground, and then walked away in a dignified manner.

Arctic had a funny feeling about his chain and post. He bit them, revolted against them, but wouldn't let us touch them. They were his. Arctic had a strong sense of property, as Trigger used to have.

The November days stayed warm. Soon we moved the wolves down into the woods on the floor of the draw, where each could choose sun or shade to lie in. On their first evening down there, as soon as Alatna had finished her supper and before the other wolves had done so, she went to the side of her chain circle and dug a shallow bed. Tundra and Killik also dug themselves comfortable bed hollows. Neither male did so.

Alatna was the gayest of the wolves, the most entertaining. One morning when first released for her free hour she hustled up and stood in the open door of the cabin, arching her neck like a proud horse but tipping it, unhorselike, this way and that, lifting a paw, her eyes bright with gaiety. Tootch lay on the cabin floor on her rug, an old down jacket, and growled. Alatna wanted that old jacket, her favorite filch, so badly. She set her teeth in the edge of it and pulled gently. I intervened. Before I could catch my wits she had my best hand towel from over the wash counter and was out the door and down the steep bank to the draw with it.

Down there she played with a blue pajama top previously stolen. She flew at Tundra, bowing and romping, raced back past Killik, giving her a whirl, ran to Barrow, jumped and fell at his feet. She caught up the towel and with him gave it the two-wolf test.

Alatna was the one the others all sought out when released. Often they would spend part of their free hour merely lying beside her. Alatna got along well both with wolves and with humans.

We had one unexpected difficulty with the wolves during these spirited days of the dreaded "hard time." It was not exactly what the conventional idea of wolves would have dictated. Our difficulty was that the wolves came into the cabin and stole things—a jacket, a hammer, the new green plastic tablecloth, couch cushions, gloves, turkish towels. Eager-eyed, wary, brimming with mischief, they slipped in the door and dashed out with their booty. Their object was not the acquisition of property but a romp, getting us to make a fuss and chase them.

Tundra was a mischievous fun-maker wolf. She was resolute and bold. One evening as I prepared dinner she slipped in and made off with a red plastic cushion from the couch. No sooner was I back in the cabin with it and had another potato peeled toward dinner than with the tail of my eye I noted the red cushion going out the door again, dangling below clear, wolf eyes, black with merriment.

On a sunny noon in mid-November the pen was ready. Alatna was off exploring but we could not wait. We led the wolves one at a time to the new gate, in the left-hand front corner of the pen. We removed choke chains and drag chains forever and turned the wolves into their new domain.

The pen lay directly in front of the cabin and plaza but lower, on the floor of the draw. It was about a hundred feet square and held as much variety as that space could contain: open sunny ground along the front; an aspen grove; dark firs along the back fence; and, chief of its attractions, two great rocks at the left and right. The one at the left was a table rock fifteen feet high and as many broad and long. A jumble of black-lichened boulders led up to the rear of it. Its prow faced a huge walrus-shaped rock across a narrow aisle. From the cabin door we could see all of the pen except its dark, secret back corners, fir-shaded. The table rock hid one, the walrus rock the other: the wolves would have a decent measure of privacy.

Wire hog-mesh fencing was set in deep trenches along the front of the pen and its down-draw side. Along its back and left side, where living rock intruded, the fence was a palisade of upright bark slabs, their bases "scribed" to follow the undulations of the rock.

The pen lacked two attractions, both fascinating to wolves and both near at hand. One was the living brook that cascaded down a steep fir-choked ravine on the mountainside directly back of the pen. The brook turned as it neared the floor of the draw and entered it down-draw from the pen. That brook would have been a great diversion for the wolves. The other lack was a far view; it existed on the ridge top above the left side of the pen, part of our acreage.

That first day in the pen it lifted our hearts to see the wolves run and play. Around and around the walrus rock they pursued each other. At last they lay belly down on top of it to rest. Even their pride touched us, as they stepped up to the water pan to drink at will, just a few licks now and then, to drink up "liberty" along with the water—liberty as compared to chains.

It meant more to them than it would have meant to dogs just to touch one another, to "troop," fur against fur, side against side.

Alatna was out alarmingly until two o'clock, when she came home jauntily, dragging her chain, and was disconcerted to behold all the other wolves loose inside the wire. I had actually to force her into the pen.

Irresistibly before going to bed that evening I went down

one more time to congratulate and "love" the wolves. Our achievement for them had been so big for us, so little for them. Having brought no flashlight, at first I could not find them. Then I discerned them, a rug of pale fur, lying under a tree by the back palisade, not wanting to be disturbed, absorbing the night and the starlight, sleeping.

I stopped. "Let them sleep," I thought, "and dream of being truly free on the sandy river bar in the Killik, of the blood-stained caribou, of the fox and her scent, and the parka squirrel among the rocks. Now they are just wolves in a pen. But the memory of the tundra hangs about them, the purpose and the promise in the daily walks—the promise of living a purposeful life, the purpose of fulfilling that promise."

Now that the wolves were together they could undertake long laborious joint projects. One was to clear away the earth from the front side of the table rock. Gradually they disclosed ledges, on which—one here, one there—they could lie in the afternoon sunshine.

Another project was digging a tunnel. They had come upon a channel of earth between rock walls under the prow of the table rock, and they excavated it, digging so far back under the rock that I feared it might tilt on them.

On top of the table rock they romped with such apparent abandon that I was concerned lest a wolf be crowded off. But they must have been guarding themselves. On the other hand, they soon learned that they dared to lunge recklessly down the all but sheer snout of the walrus rock, their paws barely touching granite.

Often they "trooped" to eat, taking their food to eat it within a foot or two of each other. (Trigger and Lady had done that during their winter chained near Point Barrow: they had carried their chunks of meat as near as they could get to the chained dog Kipi, an older animal they looked up to, to lie sociably eating while the tantalized dog grimly rose, turned and lay down with his back to the meat.) But also they sought separateness according to the circumstances, as when one wolf had a special tidbit. And Killik, nowadays, could carry away her pan to eat apart from me—not so good when the pan held milk. (Trigger used to carry his pan of milk out of the crackerbox, not spilling a drop unless I exclaimed in dismay.)

The wolves played alone as well as together. Barrow would leap and pounce, then dig as if for a vole. Once while the other wolves slept Killik stood on top of the walrus rock, in the light shadow of a bare aspen alongside, gazing upward at a raven on a branch overhead. The raven cawed indignantly down at her. Perhaps she had almost caught him, eating at the food of the sleeping wolves. She took a perpendicular leap straight upward at him. The raven hopped, then resumed cawing at her. Several times she leaped at him.

For these minutes Killik was not a captive. We were always conscious that these wolves, once free for much of the time, were not to be so again. We did what we could think of to make captivity more bearable. We led a water pipe through the back palisade, bringing a trickle of water from the plunging brook outside the fence, to fall from the end of the pipe. It projected a few feet into the pen and was a yard above the ground, so the trickle started a hollow, which the wolves enlarged to a small pool. Often they lapped there in preference to their water pans.

The wild—its scents and sounds and sights—was all around and afforded some diversion to the wolves. Homelike to them must have been the mellow kawk and toot of ravens crossing the blue; there had been ravens in their arctic home. Two great horned owls lived in the steep wooded ravine back of the pen—a male and a female, to judge by the different rhythm of their hoots. Sometimes toward dawn we were wakened by the brief sounds of the wolves rousing sleepily, apparently wakened by the hooting of the owls. Slowly the wolves' voices swelled into the full dawn chorus.

Rarely, a few coyotes sang near, their voices echoing from the granite crags. I would slip out onto the deck to listen to the wild strange choir: Hark, the feral angels sing.

One evening Tootch must have treed a cougar down by the brook. We heard a peculiar animal noise from the night and called Tootch to come indoors; she came from the direction of the noise. In the morning we found fresh cougar claw marks on the trunk of a big aspen by the brook. The wolves would have missed no detail of that nocturnal encounter.

The days had their routine highlights. In the clear gray dawns Cris went down to the pen. "There's my puppies," he called. From rocks and hideouts trooped the wolves, tails wagging. Squeaking in welcome, they met him at the gate. He knelt and hid his face in his hands and they surrounded him and climbed onto his back.

Wild beautiful play among themselves followed. Two might leap up tall and descend whirling.

I went to them at siesta time. Their bodies were plastic now. Gently I shoved them, straightened a hind leg here or there, giving it that least extra stretch of the good muscles that they could not quite give to themselves. The wolves liked this. Arctic liked to move my thumb back onto his carnassial teeth at the back of his jaws and munch it gently. While the other wolves slept, the one I was petting would tilt its head up and kiss my face.

Sociable Alatna *always* slept nearest the gate and possible visits by me. Barrow was the *only* handkerchief snatcher: he burrowed his head into pockets for a handkerchief.

At evening came another long wild play, as the last sun warmed the skyline crags and the darkening draw cooled. Now, every time we glanced out the door, we saw the wolves intently chasing one another. It made me feel good, for in my mind ached the memory of these wolves playing free on arctic snow that lifted in clouds from flying paws.

In their first big temperate-zone snowstorm the wolves looked upward with interest at the flakes. They had not done that with the driving arctic snow.

On a radiant morning of crunchy snow, Alaskan friends, a park naturalist and his wife, watched silently from the sunny deck as I led their only child, a rosy little boy of three or four, in his dark-blue silken snow suit, down into the wolf pen. I stepped away and left him standing alone in the sunshine by the front fence, watching the "big furs" with fearless interest. Dark in the morning shadow along the back palisade, the wolves leaped and crowded in their excitement over the presence of this small stranger. Their eyes were brilliant. They kept them fixed on the child, not sparing a single glance toward me. Even leaping right over one another in excitement, they kept their heads turned toward the baby boy.

But—were they becoming too gay, too excited? This was a test that could not continue. Speaking casually to the wolves and the little boy—he must not associate fear with this rare experience of the wolves' beauty—I swept him up into my arms and bore him out of the pen—and breathed! But all of us still thought the wolves would have been friendly.

In one way or another human beings as well as the wilds about us contributed to the entertainment of the captive wolves. Feeling lighthearted one cool, still, sunny day, I went into the pen skipping. To my surprise—would I ever learn that I was not all that apart from the wolves but that our moods interacted—they acted different. Tundra had always liked to have me jump at her and she would jump backward, understanding from the first time I had done this that it was only a game. Now her serious-looking spectacled eyes smiled. She jumped into the air, spreading her long-fingered paws; she jumped not directly at my face but near it. Alatna was intently nipping the seat of my jeans. The other wolves gaily jumped until my face was raked by a muddy paw, Alatna pinched me and I hurried to back against the fence. The wolves were so merry that it was fun for me.

Then Alatna followed me as I cleaned the pen. She did not care to play "jump" but had something else on her mind: she wanted to play "trophy." Denied my shovel or carton, she picked up a wing-shaped crust of wood from beside me and ran. I gratified her by running after her. The others all chased her too.

They liked a light sizable toy such as that bit of wood or an old, tooth-crushed aluminum cup that some wolf had stolen and dropped here before the pen was closed.

In spite of gaiety, the young tundra masters were downhearted sometimes; they felt their captivity. Once in a while I felt mine, not recognizing it as such. Cris had considerable mobility, driving around the country getting acquainted, and, as an innocent newcomer—for such we both were—becoming involved in a starry-eyed gold-mining project. Cripple Creek, once great gold-mining country, lay to the southeast of us, and to the northwest, near Fairplay, the county seat, a big placer-mining machine stood dormant ahead of its ineradicable trail of devastation. As for me, I stayed at home with the wolves; we did not like to leave them unguarded. Usually I did so with pleasure and good cheer.

But working along from dark one morning until noon I felt impatient to get at a new project I had—writing a book about our arctic experiences. Wanting to get at this work with a bit of my morning strength left, I felt like a person running along a station platform, dodging obstacles, trying to catch a slowly departing train. I realized that my full attention was still on "oughts": I *ought* to build a fire and boil water and cook the cranberries. I *ought* to bring up water from the brook. I *ought* to change the bedlinen. I *ought* to wash some seed wheat and put it to boil slowly. I *ought* to bake bread. I *ought* to scrub the floor. "I'm sure not brooding over my arctic book to gain a sense of structure for it," I thought.

Hotly therefore, with my lunch, instead of grabbing an anesthetic book to read while I ate, I went outdoors and sat on the edge of the low deck in the sun. "The table is spread for all, for you, too." The Steller's jay in the white aspen, the stillness and ages-formed rocks, the backlog of inner stillness—these quieted me.

Crag cabin was a good place for wolves and me. It was secluded—two miles from the only neighbor, three miles from a public road.

3 *The Yard Pen*

WE had seen almost as soon as the first, or "home," pen was built that it shut the wolves away from us as never before. Yet we–our companionship, the diversions we could provide–were the chief consolation we could offer for their lost liberty. They had to be brought into touch with us again. So we built the "yard pen," connecting the home pen and the cabin. Its back fence was the front fence of the home pen. The façade of the cabin, flanked on either side by a small gate, formed part of its front fence, on the side toward the back plaza and the roadhead.

The yard pen held little of interest in terrain–just the bare plaza and the steep bank slanting downward from it to a level strip along the home-pen fence. Halfway down this bank, to the right of the cabin, was a narrow terrace supported by a rock wall; the terrace would become a sunny lie-up for the wolves. At the left of the plaza was another sheltered spot that would be a favored lie-up, beside a rock wall at the foot of the hillside.

The day the yard pen was completed Cris drove home with the carcass of an old horse that a rancher had given him to shoot for the wolves. He spiked a crosspole high between two trees down the draw, pulleyed up the carcass and dressed it out. Then he took the lungs and a piece of hide to the wolves, not as food but as toys. They acted wary of the hide, as if it were alive—jumped back, approached. They got a lot of mileage out of that hide, their first since leaving the Killik. All tugged at once. They ran, the wolf in the lead carrying the hide.

The next morning when I went to the home pen Mr. Arctic hurried to the hide and picked it up as if to show me what they had. He shook it like a living creature. I did not know his real state of mind but this was the most valuable object the wolves had had in weeks. His eyes were black. He seemed to have some feeling connected with the excitement and pleasure of release at hand, and some feeling in connection with me, an outsider, coming into the pen, and some with that important hide. He looked proud and pleasurable. Like our first wolf Trigger, he became black-eyed and excited about property; Trigger had always cared a lot about property.

With a shout I flung the gate open. The big furs charged out past me into their new pen.

One fact escaped our attention for a while when the yard pen was completed. Our front and only door now opened into the wolf pen. We were, in effect, captives too.

The yard pen made changes for all—for the wolves, for us, and above all for Tootch. The changes for us were mainly good, but at first for Tootch the new pen worked hardship: she could no longer avoid the troop of wolves.

As for the wolves, they now had something hard for captives to come by—variety. Every morning I went down shouting and made a big thing of opening the home-pen gate. All day it stood open. Toward evening Cris, bearing a huge dishpan of chunked raw meat and followed by a leaping, tail-waving throng of wolves, led down the steep path to the home pen, fed and penned the wolves for the night.

I had something I had missed, had not known I missed, and now found continually piling excitement and massive experience into my arms. Something as free and abounding as the pure light mountain air and as easily overlooked, it was yet the one great, overwhelming fact about the wolves—the tremendous surge of their socialness, of all of them interacting among themselves and with us.

They would leap straight upward around me, their eyes fastened on something I held overhead. Sometimes it would be a dead "highway-rabbit." (Highway-killed creatures that we found were a scarce, valued addition to the wolves' diet.) The wolves would throng me at the door when I undertook to give them their daily vitamin pills. I would move my hand, holding a small raw hamburger ball containing a pill, among jaws, trying to insert one ball, one pill only, into each mouth. Killik, of course, shrinking around the outside of the troop, was the hardest one to get a pill into.

Once we came home from a visit later than usual; it was pitch dark. Over by the cabin, along the new fence, tossed a row of green sparks: the wolves were waiting for us.

I started for the cabin to build a fire. But in the yard all the wolves wanted to kiss me, except poor Killik, lurking off alone. Several times a wolf stood fore, another aft, both with forepaws on my shoulders and the front wolf licking my face. I knelt for their convenience and Mr. Barrow snuggled his head through under my arm from the rear and kissed my cheek.

All of them kissed exuberantly tonight but as a rule each kissed differently, in his or her own way. Alatna kissed casually and often. Barrow kissed lavishly whenever he could. But sometimes I thought he kissed to placate the powers over him. Travel had made him an anxious wolf. He was the most keyed up of the five, relaxing into daytime sleep the least of all; his head was always up, looking around. Alatna did not like to be disturbed when sleeping; she got up and moved away. But Barrow rather liked it.

Mr. Arctic was businesslike about kissing. He stood up, kissed briefly, then jumped down—he had attended to that. But I felt sure he liked me. Tundra was not a kisser. Usually she looked at my eyes and passed her nose about over my face. It was a rarity, a gratification when, as tonight, she vouchsafed a touch of her tongue.

These days the wolves still came into the cabin for trophies. I wanted to welcome them in; I left the door open as much of the time as possible and always thought I could foresee and prevent trophy snatching. But like a flash the wolves acted.

First though they deliberated about coming in. They hung at the door as if with delicious thrills of trepidation, the natural wariness of each heightening that of all. They peered bright-eyed into our Sesame cavern of potential trophies. One might slide his forepaw as far forward along the slick linoleum as possible while maintaining the security of his hind paws, stretched back into safety on the deck.

Or a wolf would perform a substitute action—"redirected activity." He would turn aside from the door altogether and examine the near end of the big plank table along the front of the cabin, on the deck, as if he had never seen this thing before. He might lift a forepaw and lay it inquiringly on the under shelf of the table.

Nowadays only Tundra and Alatna still entered freely, as before. Tundra was the boldest of the wolves and Alatna the most self-assured. But Alatna looked up at the ceiling a great deal and was dilatory about selecting and seizing a trophy. Tundra was decisive: she spotted and seized one in an instant. She would saunter along outdoors behind the shrinking troop at the door, as if indifferent. But when good and ready she stalked straight in—no flinching or cringing at the door. Tundra was resolute and merry.

She scratched at the closed door one day—not calling me but trying herself to open it. I let her in, coaxed, and thought she was going out. But like lightning she tipped a pile of ironed laundry from counter to floor, picked up not one but two folded Turkish towels and ran. After I got them back she scampered with her hind quarters hunched, like a cartoon of a running bear cub, her tail tucked under, a precaution against her followers seizing it. Next she dug and disappeared under the cabin and all the other wolves stood around the empty-looking hole, watching it and crying. I stamped on the floor, got her out and blocked that tunnel toward freedom.

Then she shot into the cabin and clear to the wastebasket by the back wall and out again in the same second, and all the other wolves seized her by the scruff and the fur, wherever they could get hold, as if to "taste" this great adventure of hers. The next time into the cabin she snatched the broom and bore it out dexterously.

I was learning self-control. If I really wanted a trophy back —the still camera, say—I knew I had better act casual, saunter, speak in an amused tone as if this were all a great big joke. Which it was, to the wolves. But if, caught off guard, I yelped with dismay, the wolves were briefed: they had struck pay dirt. Nothing now would have induced them to part with the trophy. Whatever this thing looked like on the outside it had to be pure platinum inside.

I used the Atalanta-and-golden-apples technique. Starting in pursuit of a trophy, I would grab the nearest valueless object, even a stick of stove wood. Making an ado over it, I ran showily but with mincing steps, not to get too far away. When sure of the trophy-bearer's eyes I threw the object. While the other wolves ran to see, the trophy-bearer would stand irresolute, wanting to go too. He might even lay the trophy down, especially if it was heavy, and go. This was my chance if I was quick and smooth and not too far away.

Sometimes as I pursued I mentally yielded the trophy to the wolves, as of more value to them for a toy than to us for a tool. Then I really put on an act. I chased and shouted and begged. Seeing me begin to flag once, Barrow whimpered. I gave up and quit and all the wolves cried. They wanted the game to go on.

We resigned ourselves to about fifteen dollars' worth of property loss a month as one part, the least part, of our project —keeping these wild-born, half wild-reared big animals proud and expectant.

The wolves had gaiety. The one that really suffered with the new pen was Tootch, the ex-lead dog. What still meant most to her was face, prestige. Now and then all the wolves gathered and courted her with unbelievable sweetness. Their tongues flipped out, kissing the air in front of her snarling face, which they dared not touch; she would have bitten their muzzles. But I thought that secretly she relished their homage.

However, the wolves could not resist mischief. A wolf took her rug, the old down jacket, one day. When I returned it, Tootch hurled herself onto the center of it, clasping it to the floor with her body. She also had meat the wolves had stolen from the birds. (We kept meat nailed to the top of a fence post and on this day the wolves had leaped and got it; Tootch had taken it away from them.) The meat, at least, she would defend. Her face looked sad and old, her eyes lightless and under a strain.

Cris himself hurt her "face" one evening. He took away her bone—one he had given, it was hers—and made her come indoors. He intended only to keep the bone from the wolves overnight and give it back to her in the morning.

The dog lay with her head on the edge of the couch and presently began grieving aloud. Suddenly she jumped up and wanted out. Soon she was at a side gate with a bone to take out of the pen—a worthless small dirty bone, but it saved her face.

Only Tundra never hurt Tootch's pride. Her devotion to the dog was a mystery. She would come into the cabin to court Tootch, who would bite her muzzle so that it dripped gore onto the linoleum. Immediately Tundra would crouch her shoulders, beam and court again. She was undefeatable. Ever since the wolf Lady had disappeared from the lives of the five puppies Tundra and she alone had made it her lifework to win Tootch's favor.

Then, as mysteriously as she had for six months courted, she won. She won suddenly. All at once one morning in mid-January Tootch was hers. The dog let her lick her face to her heart's content. Tundra seemed in seventh heaven. She stepped softly across the reclining dog and stood, as a young favorite wolf may do to an elder. She played with Tootch—a slave work-dog most of her life, a dog that seemed never to have played before.

From this day on Tootch never showed a trace of distrust toward Tundra. She gave her heart with perfect trust to this wolf. She changed now toward the other wolves and even played with them, but she never entirely trusted them.

She slept on the foot of our bed and every morning from now on, the minute she saw we were awake, she begged radiantly to be let into the yard, to be courted by Tundra. The wolf would run to greet her, bowing, her gray eyes smiling. I knelt once to share the greeting, petting both animals, and Tundra's cold nose gave my cheek a light push. She did not want me horning in on her greeting to Tootch.

The morning after Tootch's sudden conversion she started her own little game with the wolves. She went down to the home pen treasure-hunting and brought back three red bones. She insisted on tugging off my work glove and added it to her pile, on the plaza by the rock wall, at the head of the path to the home pen. Then she sat up waiting for customers and now and then scuffing dirt over her hoard with her nose. She flew at each wolf that approached. She was guarding her little "bank account."

At evening on the first day of this, as Tootch was indoors and the wolves were following Cris toward the home pen, Alatna hung back. She stepped over to see what it was that Tootch had in her pile. The next morning, Tootch not out yet, shy Killik went right to that interesting pile, all curious. She took the chief bone and ran with it but nobody followed. She carried it around a long time. It had been the most desirable thing in the world the day before.

That day I saw Tootch chasing the wolves. Somehow she had got them started to running away from her. She nearly exhausted herself among the rocks, barking as she ran. The wolves galloped easily ahead of her, or ran around the other way and met the file coming. Nearly halted with exhaustion, Tootch met one such wolf and bit it in the hind leg as it cheerfully turned to play the game and run away from her. The wolf cried out but showed no resentment.

One morning when let out the wolves, except Tundra, to our

surprise bypassed Tootch and came sweetly to kiss me. The former lead-dog did not like to be ignored. She sat with suffering ego and a look of authority beside her hoard, flying in a mean way at any wolf but Tundra that came near.

The wolves were racing below. Suddenly Tootch broke. Pleasure over dignity. She ran down and joined them. They followed her over the rocks with happy docility and easy speed. Tootch ran with the jerky elegance of a muscular dog driving herself to speed. Speed flowed out of the wolves. They ruffled and flowed over the rocks effortlessly. In reality wolves are not extremely fast animals; but they look faster than they are, perhaps because of the ruffling and flowing of their deep fur.

At noon that day Tootch came to the cabin door, panting and radiant, followed by the wolves. Followers again! Imperiously she barked to be let in. I hastened to oblige so that the dog would not lose face with the wolves.

She developed suddenly and much in the first two weeks after her conversion. She changed toward us. Beside me on the couch one evening, she put her chin trustingly and affectionately on my arm, then my shoulder, in her eyes a thirsty bright silent passion that asked to be quenched with love and reassurance.

One evening as we sat at dinner and she lay back of us on the couch, there was a sudden wail from her that caused us to swing around and gaze at her. Fixing her beautiful dark eyes on Cris, she wailed again. She could bark but sled-doglike preferred to yowl.

Cris gave me a wondering glance. Gently he questioned, "What is it, Tootch?"

At this sign of rapport the dog redoubled her strange yowls. He queried again, she replied, a conversation ensued, so extraordinary in sound that every evening after this, if the dog did not start one, Cris did.

4 *The Big Pen*

WHAT A SKY there was on the afternoon when we started building the "big pen." First it was black-blue above the sky-line crags. Next it was a phony-looking deep robin's-egg blue, between dark-green firs. Then a cloud—still to the east—turned a warm orange-buff. It reflected on the snow around us so warmly that I paused, with bark slabs on my shoulder, to look around as if in a room with gala lighting. That evening concentric rainbows haloed the perfect moon.

This big pen would be our last effort to give the wolves what they needed most—space. I longed to enclose our entire ten acres for them, but the cost of doing that was beyond us. Required, besides wolf-proof fencing, would be catch-pens or interchanges at every entrance, including the head of the road. These would each have to have two gates.

But the new pen, running down-draw from the home and yard pens like an elongated middle leaf of a three-leafed clover, would enclose far more variety than the wolves now had, as well as space, another acre or two. It would hold the brook at last, with all its changes, hooded under woods and thickets. Along the right side of these ran a long strip of open ground that we called the "avenue" and that would make a fine raceway for the wolves.

On the left of the brook a rugged wooded hillside climbed to the bench. Looking diagonally upward across the draw from the plaza, we could see only the near end of the bench. It sloped quickly away and out of view, first through open sunny space, then an aspen grove, to dark fir woods at the far end, part of the native forest outside the pen. The wolves would have much seclusion and even some secret places where they could hide.

They would have granite towers as lookouts. One at the head of the steep trail from the brook up to the near end of the bench had a sheer face, which made of the spot at its foot a good lie-up–shaded part of the day, and always a "location," a "place." The wolves would like to lie there, backed by the rock and commanding an overlook of the plaza and cabin and the activities around them.

Near this "pen tower" was the "fence tower," forming part of the fence, though most of its fall–forty feet perhaps–was outside the pen. It plunged downward into the steep ravine. But its top, inside the pen, was low, climbable by easy ledges. The very top was an open slant and I was afraid a wolf might be crowded off here, especially when the slant was snowy or icy. Eventually wire fencing was strung around the drop-off to guard against disaster.

The fence along the front of the big pen was to be a palisade of upright slabs, like the one in the home pen. The ground was frozen now. Every twilight, down the snowy draw against the backdrop of black firs, glowed piles of red coals, each pile under its waver of thin silken blue smoke, thawing the spots where, the next day, postholes would be dug.

At last came an evening when we knew the big pen would be ready by noon the next day. I ran down to the home pen to share my joy with the wolves. I threw in a pasteboard carton to keep them from the gate while I slipped in. Alatna seized it and made off. I chased her happily. Arctic picked up a piece of meat and stood near her as she tore up the carton. He cried; he wanted someone to chase him. So I did. Then Killik got herself a piece of meat and watched me! I really gave Killik a play.

Next, Barrow tugged at my glove. I let him drag it off and when he got up onto the rocks with it, I stood below, reached up and patted at his paw. He was in a dilemma: he wanted to kiss me and he wanted to hold onto his glove. He lifted an "arm" sidewise as if to pat at me. He tilted his head and leaned down. He always did like to stand above, bend his head and kiss me. He did and the glove tumbled between me and the rock.

In the morning all was gaiety. The wolves seemed to sense our excitement. To entertain them during these last hours while the big pen was being completed, we tipped out a carton of new bones on the edge of the plaza. Tootch immediately took charge. Alatna came up, followed by the rest, and when Tootch jumped at her, Alatna just fell over and rolled belly up. What fun for the wolves! Tootch was taking the bones importantly. While she chased a "thief" below the plaza, Arctic came up and took a big chunk of bone. Did he run with it? No! The big fellow stood there holding his head low, his jaws wide over the bone, watching until Tootch came up the bank. Then he ran. But she did not follow. So back up the path he came with his bone, to make sure she saw him. This time she chased him.

At noon, coming in the side gate with beets and apples for lunch, I was mobbed by the wolves. They knocked the gate wide. I let out a screech that fanned them down the bank, looking over their shoulders, with liberty at their heels. Old Arctic got the biggest beet and waited down at the home pen gate to see if I would chase for it. The bubbling-shower feeling of exhilaration!

As we ate lunch, Alatna, beside us, was licking up a bowl of eggs. The screened side panels of the picture window stood open, the door was open too. A gust blew papers from the table, a magazine flapped open, the door started to whiff to. Whirling to flee, Alatna lost control, her paws shot from under her on the waxed linoleum. Flat on her belly, paws oaring to all sides, she snaked out the door.

Having saved herself, she stopped and gazed back at the door, now propped open, as if thinking, "I always knew that place was going to be dangerous someday."

The finishing touches on the big pen were put on the small new gate down in the back right-hand corner of the yard, so the wolves were at hand and all excited.

The gate was opened. Tundra was the first one in because she was nearest. Alatna was next, then Killik and Barrow. Arctic was last. He studied the overhead framework of the gate, his eyes black-bright with excitement. He tried several times to go under it but recoiled, like a new diver. At last, crouched low, he shot through.

Ecstatic running! Gray forms were shooting up and down the steep hillside, disappearing onto the bench or behind the big rocks and trees.

Alatna came to me, her ears tucked back, simply crying from the pressure of feeling. But sadness touched me: I knew the wolves thought, for these first minutes, that this was freedom. They would be finding the fences.

Our plan was to pen them nightly as usual in the maximum-security home pen, but after a few days their pleasure in the new domain defeated the plan. They chose rather to go hungry than give up the big pen. So both gates were propped open; the wolves had the run of all three pens.

They had considerable hunting to do in the big pen. Perhaps a blind feeling that there would be dawn and night hunting to do there had hindered their submitting to being penned away overnight. Golden-mantled ground squirrels pursued habitual routes, passing easily through cracks in the palisade or the hogwire-meshed wire. They had hiding places at hand, recesses going back under rock ledges, where, watched, they were nevertheless secure. Cottontails came into the pen. They too had a good place to hide—a big brush and pole pile near the brook. The wolves dragged at it but enough remained, or was replaced by us, to hide the rabbits.

These days the wolves had so much excitement and pleasure that all except faithful Tundra often ignored Tootch and her little pile of treasures on the plaza. Finally the ex-lead dog hit on a way to get into the picture again: she carried her hoard one piece at a time down to the new gate and sat there guarding it and snatching authoritatively at each wolf trotting through.

Wolves are not brusque, simple animals; they are intricately social. Among our wolves obscure changes were taking place. I could not read them.

Preparing dinner one evening I tossed the stem end of a squash to the wolves for a toy. Alatna got it. As usual I chased, to add to the fun. Alatna slipped and dropped the toy and could not pick it up in time to run on from me, so she brazened it out—lay with the squash end between her forepaws. I had been shouting to increase the gaiety, so the wolves were all excited in the chase.

Now came Mr. Arctic, whining with I knew not what feeling; he stood beside Alatna, looking at me. But this was not the game; it was something new. The whine was new, the action was new. I felt mystified. He wanted some course of conduct from me, but what? Baffled, I stood still, questioning him. Then, uncertainly, I withdrew. But was this what he wanted? I wished I had a Rosetta stone to translate wolf into human communication. I should, in due time, receive the Rosetta stone, or a fragment thereof. Meanwhile I was tractable material: I skirted danger unaware.

Another time quick Alatna was again the wolf that got and ran with some trifle I gave. This time I was busy and did not chase but went down a few minutes later to see why the chase by the other wolves had ended so abruptly. Alatna lay in the edge of the woods, gnawing on her tidbit. Mr. Arctic stood beside her and bit or menaced any other wolf that came up to investigate Alatna's trifle. He offered no interest in it himself.

Those two seemed paired. If any wolves were to be found apart, by themselves, it was Alatna and Arctic. Ever since puppyhood they had been pals, but they seemed to be closer now.

If this really was a change hinting at their first mating season, still almost a year away, would their closeness last until then? Or might Killik prove to be a "dark horse," supplanting Alatna at the last minute? It did not occur to me that siblings might refuse to mate.

Another change among the wolves was that Arctic and Barrow were showing a new authority toward me. One day Alatna tried to take a carton away from me. I needed it and held on. Barrow and Arctic took Alatna's part and jumped on me. Recognizing the situation as a contest, they had approached and watched with interest. Barrow grabbed at the seat of my jeans, Arctic gripped my elbow in his teeth. These were their respective ways of disciplining me. They never exchanged methods; each wolf stuck with his own way.

The wolves were fond of a stew I had devised. We had no idea of how to feed wolves in captivity, but if vitamins were good for people, perhaps they were for wolves also. So I ground up either beets or carrots, added a bit of onion, sometimes an experimental bit of raw cabbage or apple, and pressure-cooked the mixture for one click only of the pressure gauge. I added oil—butter or ham grease—and served the stew warm in big aluminum bowls, set in a long wooden tray. This stew was not a main dish but an added tidbit.

One evening some stew had spilled behind the bowls and the wolves were still trying to get at it when I picked up their tray and started toward the cabin. The wolves did not like this. I felt Barrow at the seat of my jeans.

Clearing up the yard pen another day I picked up some property assembled by Arctic, the property lover, on the terrace below the plaza. I should have known better. My first baby sentence, so I had been told—"The Belle she did!"—had been a protest to my mother over a like invasion of rights: the hired girl Belle had cleared away my toys. Arctic seized my elbow in his teeth.

I remembered that just a year before this Trigger and Lady had changed toward us and had begun to enforce proper discipline on us. Wolves have a strong sense of "ought." Once on a long tundra walk, as Trigger had lain down to rest, Cris had knelt teasingly and domineeringly over him. Trigger, his eyes black, his lips drawn back, had laid a forepaw heavy on Cris's arm. I had sensed then that we needed to change our ways toward the wolves. "He's a—he's more like a 'man' now," I had stammered, trying to convey my intuition. "He needs to be treated with respect."

But Trigger and Lady had been able to leave us at will; their great free life on the tundra had obscured their change; I had forgotten about it until now. So once again we had made the mistake of supposing that the youthful submissiveness of wolves was "wolf," whereas it had been merely "puppy wolf." Now our five wolves were maturing psychologically: they were "turning a corner." Half in the dark about the changes in them, would I be able to turn the corner along with them?

Another of the changes in the wolves was that they were increasingly jealous. All the wolves we had known had been extremely jealous. Even the totally wild wolf that we called Silvermane had been jealous of Lady and of me, over Trigger. But nowadays the five seemed more jealous than before. These days they took their siesta over on the bench in the bare aspen grove. Going to them at siesta time one afternoon, I came first to Tundra and Alatna, lying near each other, Tundra nearer to me. Alatna began to stretch and curl, expecting caresses. Undecidedly I paused by Tundra. Alatna got up and stalked away!

Once as I knelt in the moonlight petting Arctic's big head, and Barrow came up to kiss me too, Arctic jumped on him and they fought—merely a moment's squabble but unusual. Another time Arctic came right up and jumped on Barrow, who had joined Alatna and me. Even an angel could hardly have trodden among the delicate sensibilities of these wolves without bruising some.

Could wild wolves afford jealousy like this? I thought that, with them, surplus emotion would be sopped up by the hardships of their wild life. Besides, they would probably have certain strong emotions denied to our captive wolves—big absorbing feelings such as that connected with the pleasurable ceremonies, described by Adolph Murie, of wolves preparing to depart for the night's hunt; or the mutual self-congratulations of a pack discovering big potential prey ahead.

In our five young captive tundra masters, blind hunger for their birthright of travel and adventure and achievement must have gnawed at their nerves. Often they wearied of captivity and were restless and bored. We knew too well what they needed. Only a year before, in the North, we had witnessed two hunting incidents of a kind lacking to our wolves.

We had walked for eight hours one day and come upon two wolves ahead, on a promontory, so intently watching something at its base that they were oblivious of us. Stealthily we moved to see what it was. On a frozen pond below lay fourteen cow caribou.

The wolves started down the side of the promontory toward them, not crouching nor stalking but moving along easily and quietly. There was no cover. Not until they were within a hundred feet did the caribou notice them. With a frantic scramble of skidding hoofs the caribou were up and off the ice and running, speeding upward over the vast uplands toward the white bases of the mountains. The tan wolves lined out after the tan caribou. As they all grew small with distance, the space widened between wolves and caribou. The wolves stopped. The caribou sped upward and away.

On a later occasion, as we ascended the point of a snowy ridge, just as our heads rose over its top we paused. Ahead, two wolves were watching caribou just coming up from the far side of a parallel ridge onto its top. About forty in number, they stopped to graze in a wind-bared hollow on the near side of their ridgetop.

The wolves moved downward into the narrow valley between the two ridges, holding their direction though out of sight of the caribou. Nearing the top of the opposite ridge the wolves, to my surprise, abruptly veered. They entered a gully I had not noticed. It crossed the ridgetop to the left of the grazing caribou. Not until the wolves were back of the caribou did they disclose themselves.

They sprang to the chase. The caribou fled down the ridge side, running our way. Gathered into a dense bunch, swerving as a unit, they swept past below us with a rush-crush of hoofs in snow, and white ptarmigan rising in a V ahead of them. Then the caribou were past and gone, on their migration direction. In the valley the wolves stopped, completely and hopelessly outdistanced.

Those two hunts had failed. Wolves fail far more often than they can succeed. But these wolves had for these minutes put forth their full powers of body and "mind." They had lived to their fullest. It was rich total experiences like these that our wolves must blindly have craved but that were denied to them.

Spring was making them restless. Two springtimes come in the Tarryalls. The second comes with dramatic suddenness, almost overnight, in late May. It is the leafing out of the aspens. The first spring is obscure; one has to look in order to see it. Bits of green herbs show; tiny fringes of green edge the dead-looking bushes. Like small tulips, blue anemones, still clasped in their leaves, crouch low to the ground, to lengthen upward later. Up on dry ridges pincushion cacti bulge the granite grit and open improbable pink-silk flowers. But, over all, the land still looks dead; the aspen groves are dun-colored mists on the mountainsides.

This first spring was with us now. Its gift was occasional days of wild sweet perfume, not diffused all over but in spots. Out walking with Tootch, I would step into an invisible well of perfume, its source utterly unfindable, then out of it.

Our greenhouse was a tide of green—pungent-smelling young cucumbers, Bibb lettuce, young cabbages, radishes, parsley, turnip greens, small bright-red tangy tomatoes hanging like globes on their tall aromatic plants. Our salads were excellent. Down in the pens the wolves nibbled the new green shoots of wild grasses.

I went to the big pen one morning, as usual on rising, to say good morning to the wolves. Only Arctic and Alatna were in the pen. The other wolves were gone.

Arctic and Alatna were leaping up on their hind legs, bickering and snatching at each other from nervous tension. The others must have gone out when these two were off by themselves and they had not been able to find the escape place.

It was a sad busy morning. We, too, looked vainly for the escape place. We got Alatna and Arctic into the home pen and closed both small gates and opened the big drive-in gate at the far end of the avenue. The truants would have to come home of themselves. It was useless to look for them. They could have left hours before and be miles away by now, and in any direction. Besides, we had not the slightest means of compulsion even if we should search and find them. Cris began putting up our first overhang wire, around the top of the home-pen fence.

The three wolves came home at noon, happy and exhausted. We closed the big gate and opened the small ones. Alatna rushed out to greet the others. They were glad to be together again, they trooped, they were demonstrative toward me. But Arctic, terrified by Cris's working in the pen, was afraid to leave the place.

After a while I went in and dragged him out by his forepaws. The big fellow dipped his cold nose tip to my bare hands and whimpered. Cris, watching, said the wolf's tail was clamped to his belly and that he urinated a streak as I dragged him on his hind legs to the big-pen gate. I let him go. We gave him hurrahs as he galloped through the gate.

Afterward when I went to the big pen he was determined to kiss me, to stand up, his forepaws on my shoulders, and lick my face. I dodged; my face was tender from sunburn. But successfully, carefully Arctic licked it. Then he jumped down decisively. I felt impressed and touched: usually he did not bother to kiss me nowadays. Was this kissing due, in part at least, to my helping him to get out and join the other wolves after he had wretchedly trotted the fence, alone in the home pen, for a long time?

Killik had brought home trouble—a nose full of porcupine quills. Neither of the other truants had a single quill. This meant that they must have stood back with wolf wariness and observed while Killik approached the unfamiliar little animal. It also meant that Killik, away from me, had been her old aggressive, confident self, as when a puppy.

She lay welcoming and sunny and passive as I went to her. This was different from any way she had been toward me since the near-strangling I had given her. Of course freedom gives trust, and she had just had a big dose of freedom. I felt that she wanted me to help her but it did not occur to me that I could do anything for her directly.

I gave her sleeping pills. She was apparently unaffected by them. I gave her two more. She stumbled and staggered but wobbled happily along with her playing fellows.

The next morning, half sick at the stomach with despair, I took the pliers and went to Killik. To my profound astonishment, she lay still as I yanked out quills. There was no time to spare on my feelings; I worked rapidly. Once she got up and moved away but soon lay down among some bushes, and let me work on her again. Trying for one bad quill, sunk to the hilt in her black nose tip, I tried kneeling astride her, to hold her down. She turned clear eyes upward, looking into mine as if to determine what I meant by this, as, when I had half strangled her, her clear thoughtful eyes had searched mine, to understand. I did not get that deep quill but the rest I did get.

The next time I went down to the big pen Killik trotted straight and quickly toward me, along the avenue. I knelt, not knowing what to expect. Killik kissed me!

From now on, she kissed me freely and often, and I bore her kisses stoically: they hurt. She turned out to be a terrific grooming nibbler. Alatna would kiss quite a while before I felt a nip coming on, but Killik went right to work—kiss, nibble; kiss, nibble—before trotting off.

At twilight on Wednesday, May 4, I went to the pen to say good night to the wolves as usual. They were not playing but standing silent, just beyond the gate to the big pen. Arctic was downcast. He wanted to bite my jeans, a new pair worn for the first time. I entreated and waved my hand in front of his nose. He desisted but as a substitute demanded my work glove. He dropped it at once. He growled at the other wolves. He even bit Alatna's muzzle lightly, holding it in his. She stood still, crying. That temperateness of wolves.

Arctic's lips drooped. That made me feel bad. I noticed it as he started for the path to cross the brook but stopped and stood, just sad—no heart in him. The wolves were always emotional, swinging from gaiety to sadness.

That night was the first of the full moon. About eleven o'clock we heard Arctic howl several times, so sad a sound. At the end of one howl there was a broken upturn of voice like a questioning sob.

We heard a noise out toward the road and, looking, saw the wolves moving in the moonlight among the trees beyond the road. They went up the side of the ridge and away, Arctic with them, having found the escape dig.

We were not much troubled. They had always come back to our camp in the Killik. Recently the three truants had come home here at Crag cabin. We thought they would take a holiday for a few days and then come home as always. Then we remembered the Herefords. The pastures out along Tarryall creek were littered with newborn daisy-white-and-red calves, lying near their mothers.

Cris left at dawn to warn ranchers and inquire whether the wolves had been seen. I stayed at the cabin with Tootch, all gates open, to receive and pen the wolves if they came, though we did not expect them so soon.

"They were made for happiness," I thought. "And for work—for 'projects.' I have been a jailer of those I love." My thoughts ran: that every man was out against them; against all wolves. That they were so well disposed, fearless, toward all strangers. For so short a time, a day perhaps, their cups would run over with happiness. "I want them killed easy," I thought, "–if can be." Quick out, not slow. For five months they had been free much of the time in the Killik; for seven months captives.

All that day there was a hullabaloo in the outside world. I was apart from it, miles from a telephone or a public road. Returning at evening, Cris said that at 2 A.M. the wolves had played in the full moon with an old dog; Rip, at the nearest ranch upriver, some miles to the west of our place. But by noon they had been miles to the east of there. They had slipped out of the forest and stood at its edge, silently watching a ranger who sat eating his lunch on a log. They would have expected him to toss them bits of his food. We had always shared our trail lunches with them. Wolves seem to expect the same general conduct from animals of a kind, unless taught otherwise.

Would the five have hurt the ranger? No. They would not molest human beings. As certainly as hunger or sunrise we knew that to be so. They would not have gone too close to the man, nor permitted him or any other stranger to approach them too closely, but they were not afraid to be seen; they were friendly and interested toward everyone. They did not know there were enemies among them.

We did not think about it, but already they had reversed direction, were going eastward instead of west. Had these tundra wolves felt the lift of the continent under their feet or seen its rise ahead? Did their nerves whisper, "Down means out–out onto the tundra, where in the spring prey might be found"?

The next morning, Friday, as Cris was about to drive away to seek and inquire, as he would do daily now, I stood silent for a minute by the open window of the cab. In a low voice I said, "It's hard to think they're 'varmint' now. Tundra and Alatna varmint!"

"Yes." He drove away.

Again I was alone with Tootch and the open gates and my thoughts. Tootch was so lonely that she would not eat; she had not eaten since the wolves left. "She's a sweet little thing," I thought. "No substitute for the wolves though." I could not name it, but to go to them in the pen was like going to health. "Clear, uncomplicated, they're your *equals.* They have no doggish cuteness but great sweetness. They're not depending, seeking, asking, needing. They're free, clear, intelligent—strong and on their own inside. Friends."

Cris brought word that evening that the wolves had stayed for an hour or so near a man who was getting out logs in the forest, several miles to the east of where the ranger had seen them the day before.

Early on Saturday morning a woman doing ranch chores saw them in a pasture only seventy-five yards from cows and calves. "I didn't get a gun," she said. "They weren't bothering the cattle. They were playing among themselves. They played so pretty I just stood and watched."

That night Arctic was shot. As some young fellows ate and drank in a lighted garage after a dance at Florissant, a village east of Lake George, the five had appeared silently at the open door and watched them, hoping very much by now, no doubt, that bites of the food would be tossed their way. One of the men shot Arctic point blank through the chest with a .22 pistol. Arctic and another wolf, also wounded they thought, fled in one direction, the rest in another.

After this the tracks of only three wolves were found. Snow fell and for a couple of days tracking was easy. The tracks moved always eastward, along the Platte River watershed toward the fall of the mountains downward to the high plains.

Arctic's death would be slow, taking perhaps a couple of weeks. I made posters offering a $50 reward for each wolf killed, and we had them posted in country stores out along the main highway. We hoped the offer would induce people to shoot to kill, not wound. It was all we could do for our wolves now.

We still thought those that were able would come home. Topping the ridge with Tootch one afternoon I saw four gray animals running in the pasture below. Immeasurable relief flooded me. "I always knew they would come home if they could." The animals were bighorns.

Rumors and tall stories circulated. One rumor was so unwolflike in every detail that we looked into it and were confounded to learn the mustard seed of fact it was based on. The rumor—circulated at Florissant, not at Lake George—was that the wolves had gone right into Lake George and knocked down a little girl and scratched her face "something awful." We found that I was the "little girl"! The grain of fact was that I had appeared at the Lake once with a trifling scratch on my cheek and had explained that it was caused by a wolf jumping affectionately to kiss me.

We could smile at the tall stories made up for fun, not hate. The wolves had been seen at the airport near Colorado Springs, inquiring about one-way tickets to Alaska. They had hijacked a jet from the field, else how could they have killed a cow at one end of the state and another, half an hour later, at the other end of the state? Most cows and calves dying that month of the ailments common in birth and infancy were "wolf kills."

But at no time did the wolves actually molest any livestock nor, I would guess, any wild animal larger than a rabbit. They were wary of big animals and had never learned to hunt. Given time they would doubtless learn by themselves. Trigger and Lady had almost certainly learned from wild wolves, while safely based with us. In Alaska the five had actually killed a caribou, but it had been a sick animal, all but suffocated by cysts in its lungs; it had lain down of its own accord, untouched. Even then the wolves had hesitated to approach it.

A week passed. More clearly than ever now I saw that what we had given the wolves had been captivity, not home. The gate in the Arctic had always opened; the gate here had never opened. I had known all along that this was captivity, but Trigger and Lady's always-returns had blinded me.

Constantly at the back of my mind I heard a far-off wolf mourning howl. It ceased one day and I did not hear it again.

5 *Alatna Alone*

About two o'clock on the morning of May 18 both of us were wakeful. It was storming. A tree down the draw that creaked against another in storms was "howling."

A deep mournful voice bayed right at the door. Home! I began calling and fumbling on clothes in the dark.

It was snowy outdoors. The wolf came right to me on the plaza in front of the cabin, crying to me, lying down, kissing and kissing me. I could not tell at first which wolf it was, its coat was so dark with wet. It was Alatna.

Tootch ran eagerly to her. The wolf was not Tundra. Tootch bit her as formerly and left her. For the dog's sake I almost wished it had been Tundra that returned.

Alatna was jumpy about Cris. Whenever he moved, she shadowed into the trees. He went to close the big gate at the far end of the avenue and at last I thought to close the side gates by the cabin. Alatna was now in the big pen though. Each time I went to her she came up and lay down under pressure of feeling. She cried.

I dragged her by the front paws into the yard and shut the gate to the big pen. I ran for food. I found her in the home pen when I went out. She ate hungrily. Then we built a fire and thawed beef heart and horse meat. She ate those things quickly. But she was still so emotional, crying and "talking," that we gave her a howl, hoping it would calm her down.

Cris backtracked her by flashlight in the falling snow to where the road was melty and he could not track. She had followed the road.

He left about 6:30 the next morning for word of whether the other wolves had been shot on their way homeward with her. There was no word. They had not been seen.

We learned their fates eventually. Lost, panic-stricken, bewildered, they had died out on the high plains far to the east, each one alone, far from the others. Killik was shot miles to the southeast of Colorado Springs, Barrow seventy miles north of there, and Tundra a hundred miles by straight line to the northeast of where Barrow died. She had been within ten miles of the Nebraska state line.

They would have had to cross the main north and south highways along the foot of the mountains. I thought that if a car rushed between them as they started to cross, any wolf ahead would flee onward, the others turn back; they would never have rejoined: they had no rendezvous points; cars would obliterate scent on pavements.

Barrow had been the first to die. When friends in Alaska saw in their papers the picture of his face as he hung head downward beside the broad, stupid, dimly complacent face of the fellow who shot him, they felt concern for our wolves.

Tundra had been the last to die. She was 180 miles from home by straight line. She gave herself up. That was like her: she was always bold, resolute and decisive. Starved down to her puppy weight of the previous fall, doubtless lonely beyond understanding, thirsty too perhaps–that is forbidding, desolate land to look down upon from a plane–she had come to a farmer plowing in the field with a small cat. He might have seemed a bit homelike to her, like Cris plowing his garden. She had stayed near him.

The man left the cat and in his car drove two miles to his

home. Tundra followed along. She waited in the yard while he went into his house, got a gun and came out and shot her.

Alatna had to be comforted. She grieved and we began a campaign to comfort her. This animal had never been alone in her life before, not even in the uterus. We felt sure that she was the one that had fled with Arctic on the night he was shot and that she had stayed with him until he died and then had come home, expecting to find the others here. She could not know they were dead. Our belief was confirmed later when, tearless, I identified Killik's skin—by the bump in her nose of the quill I had not got, and Tundra's—by a scrape of the guard hairs, made as she had dragged out between granite nubbles and bark slab, escaping once.

Alatna did not grieve silently like a cowed wolf. Solemnly her great wolf mourning howl reverberated through the draw, from mountainside and crags. There are no adequate words to describe wolf sounds. The words we do use are tainted with human fallout. "Howl," "whine," "whimper"—these are contaminated with irrelevant connotations.

The mourning howl was a minor rise and fall, repeated over and over. It might rise from D flat to G, return to D flat and sink away to C. Or go from D up to G flat, come back to D and die on D flat. Or it might rise only from G flat to A, return and die away on F. Always it ended in this "dying fall," a semitone lower, like the wail of a grieving woman. The tone was pure unless, occasionally, it blasted into a mere yell as if from a sudden access of feeling. Its import was unbearable; we had to help Alatna.

But could comfort cross species lines, from human to wolf? We did not stop to ask. Impulsively, on the spur of each new sign of need, we acted.

The first day was not too bad. I had expected her to mourn-howl. But when I looked out she was walking silently by the home-pen fence. She ate breakfast hungrily. Afterward I stayed in the pen with her until she fell asleep. Tootch lay near. She had hunted and got a bit of hide to guard.

I went down a time or two to see if Alatna was awake. She still slept, wet and grimy, under the overhang of the table rock. When I found her awake, she laid her head on my arm without getting up. She kissed me and yawned. I was glad to see that. In the night she had been too keyed up to yawn. She "talked" to me some more—her emotions and experiences were still a pressure in her. Then she slept again, so ready to sleep that I thought she must b- letting go after long vigilance.

At one in the afternoon it was still snowing, she was still sleeping. This was the biggest snowstorm we had yet seen in the Tarryalls. The sense of its approach might have speeded Alatna's return.

Within two days we had set the pattern of our campaign to comfort her, not realizing that we had a pattern nor even what it was that we were trying to guard. But wordlessly we felt a value here, such that anyone would spend his strength to guard as a matter of course. Our campaign took three forms: we entertained and diverted Alatna; we yielded to her, not imposing our will on her; and we tried to include her in our human family.

First off, trying to entertain her, I took her a deerskin that a man had traded to us. She gathered up the end I had, folding it to her. She stood on it and whimpered, protesting, then lay down and gnawed it. I yielded and let her have it.

Later I took up an old sun-greasy bone and ran with it. When she got hold of it, too, and I held on, she saw I could get it away from her. She protested again by a complaint cry. Once more I gave up, as wolves often do when adjured, and walked away.

She controlled me by force as well as by voice. She pushed me away with her nose, to leave her alone in her new hay nest out of the rain, under the overhang of the table rock. She pushed—barely—my crowding hand, where she lay gnawing unhungrily at a Wyoming ground squirrel Cris had shot and given to her. Also she still controlled by crying a little to protest, as if to say, "Desist; I'm unhappy about this," when otherwise she would have had to use force to change my actions.

Neither of us went to bed without stepping outdoors and calling good night to the lonely wolf in the darkness down under the big twin spruces near the gate in the big pen—her lie-up in dry weather. She was silent but she heard, we knew that.

So this was our pattern: we diverted her, yielded to her, and crowded our human family upon her.

I never let her go away frustrated from a closed door: I opened it every time she came, though that was scores of times a day. Even if the table was freshly set for lunch, I cleared it, to wash and reset it later. Nothing ever stayed clean long.

If I could I dropped my work and played with her when she came indoors. At least I talked to her. It is a pleasure to talk to a wolf. Its voice is so like one's own that all one's tones and inflections seem to be noted.

Often I tried to give her a toy or tidbit. If there was nothing, absolutely nothing, I thought, "There must be *some*thing." Generally I found it. My ingenuity aroused my surprise and admiration!

Cris played "jump" with her whenever he came up from the garden. With a smile and a handclap he made tiny electric jumps toward her. She wagged her tail and spread her forelegs wide, crouching her chest to the ground. At every jump she tensed galvanically and tossed her head, looking mischievously aside. Suddenly she would spring away into a big run, circling back to start the game again.

Our friend Carol visited us. At twilight on the first evening of her stay she followed me down to the home pen to meet Alatna. In a low courteous voice, "Alatna," she said. The wolf made to play with her—flicked back her ears for an instant, a revealing, thrilling mark of friendliness, then snatched up a toy, a stick, and ran. The game of trophy! Carol was the only person but ourselves with whom Alatna ever offered to play.

Unlike human grieving, stabilized by words, Alatna's could be distracted. She could feel gaiety. But the minute distraction ceased, off toward the bench she headed and soon from the woods at its far end resounded the mourning howl. Then if I could leave my work I went down into the draw and across the brook and up the steep handhold path, beaten by wolf feet, to the bench. I walked along it, hunting for her. She would stand silent at my approach and I had to look behind towers and rocks and trees. When I found her I played with her. At the same time I talked to her: endlessly I talked to her, exhorting, upbraiding, consoling.

If I could not leave my work, at least I would stand on the deck and shout encouragement and remonstrance to the unseen wolf in the far woods. I knew my voice carried to her across the treetops in the draw. But her mourning howl floated to me, unchanged.

One day, at my wit's end, I shouted absurdly, "What *is* it, Alatna? What do you want?"

Silence. Then she spoke in a voice that was totally different: not the big absorbed grieving howl, but a tone simple, tentative, inquiring. I queried again, she replied, we conversed across the length and breadth of the draw, I down on the deck in the sunshine, Alatna off in the dark forest. Daily from now on we conversed in this way.

We kept a fork for Alatna beside our own at breakfast and from its tip she took bits of hotcake or bacon. But the door had to be open behind her, and every few minutes she stepped outside and checked up on the surroundings.

Some things I did, others I avoided, all for our nameless goal. I know now that the goal was Alatna's autonomy, her sense of having some control. I know, too, that we valued and were trying to preserve more than that: the autonomy of wild animals gives a freshness and release to human beings.

I yielded to her autonomy in small things. If she wanted to hold her tongue glued dreamily to the satin of her stainless steel bowl I did not take it away from her, I knelt holding it until she released it. If, after a meal, she reached her head for one more inspection of the departing bowl, to make sure nothing had been overlooked, I returned it for her to look.

In the cabin I never walked quickly toward her, causing her to quail. I slowed as I neared, and lowered my height a little, as wolves do when showing submissiveness toward an animal they court. On the other hand, it actually helped her confidence in the long run for me to walk briskly past her, my eyes on something I was going to get. She shrank but took note that I acted objectively, not with reference to her.

Without reasoning it out, I began to put predictability into her world: I told her what would happen next. Predictability is a form of control. If the big-pen gate was closed, on going to open it I would sing out, "Let's open the gate." Then, "Gate's open." She learned what would follow these phrases, so when unexpected company trapped her in the closed yard pen, instead of panicking and throwing herself at the fences, she would walk quietly to the gate on hearing me call the first phrase. Detaching myself abruptly from the guests, I ran to fulfill the second and she trotted away into the decent privacy of the big pen.

Like Arctic, she demanded to examine everything carried in or out of the cabin and we let her do so. However, if it was something that would not interest a wolf, I said disparagingly, "It's nothing, Alatna." She verified this many times, came to believe it and did not approach to look—a trust I traded on when what I brought was meat for our dinner.

Aside from that, I was scrupulously, perhaps ridiculously, honest with her. I could not bring myself to say, in comforting her, "It's all right, Alatna." All I could truthfully say was, "It's not so bad." Perhaps her wolf intelligence gleaned the honesty not of the words but of my concern for her.

She always responded to sincere tones, never to phony fawning. I would apologize for mishaps, such as a stepped-on tail or paw, as civil people spontaneously do to an animal they have hurt unintentionally. "It was accident, Alatna," I would deplore, kneeling beside her. (I omitted the article as if Indian brevity might make the meaning clearer!) Evn if I was stifling laughter, my regret was sincere.

She always accepted apologies, not by any special act but by her air of heightened cheerfulness as she stepped around.

Most of our actions were impromptu, but there were a few things that I deliberately thought up. She was wild with impatience to follow when I went out the side gate and toward the storeroom. She would leap along the fence, lifting herself so high horizontally that I thought she might sometime make it and shoot out over the top. But that was not the reason for what I did: I would whine at her complainingly, "Now don't, Alatna. You stay there," as if she had an option. I hoped my peevish tone made her feel that she did.

Slowly I became aware of a fact that struck me as curious: she seemed to expect some role of me and I was not playing it. But what was it? I thought this over and decided that perhaps she wanted to be number one animal in the pens, so I willingly tried to enact the role of number two animal. I have no idea by what antics I tried to convey this improbable notion, but whatever my histrionics, they were not, in her eyes, Academy-Award stuff. She would retire and observe them dubiously.

At last it dawned on me: what the realistic wolf wanted was for me to be myself. I was, de facto, number one animal in the pens. I accepted my proper role and she seemed more contented.

Another intangible was that she seemed more contented when Cris was around, his hammer resounding, the chain saw or cat buzzing, our cheerful voices calling. Perhaps the "establishment," shorn as it was, seemed more rightly complete when both of us were there.

I thought our cheerfulness and frequent laughter made the emotional climate brighter for her. The cheer was genuine, not affected. In spite of grief that recurred for our lost animals, we had new projects that absorbed us—a garden, the writing of my arctic book, happily started at last.

The hardest thing to provide for Alatna and what she needed most, next to wolf companions, was a sense of achievement. I tried to think of how to give it. Letting her kill an animal would have helped. But we could not afford a calf nor obtain a sheep. Besides, I could not have turned a helpless big animal into the pens. What I dreamed up was so puny compared with her need that it was ridiculous, but I felt proud of it and thought it helped. I would give tidbits not from my bare hand but in a plastic sack. This Alatna could carry off as a trophy and tear up at her leisure, discovering the contents. I liked to see her trotting off down the avenue, busy and interested, with her little sack.

Which of the things we did really helped? Did any one thing help, or merely the sum of them? Or our concern? We weren't gaining much but we weren't losing either.

The lonely dog might have helped the lonely wolf by playing with her, but she would not or rather could not. Alatna courted her sweetly but her heart was not in the courting as Tundra's had been and Tootch knew the difference.

In estrus, Tootch escaped from our pickup at old Rip's home, upriver. The two dogs eloped to the hills. Cris drove over daily to inquire whether they had returned. Rabbit hunters were numerous and might shoot them. From the side gate Alatna watched the pickup return. No Tootch. She stayed at the gate, looking down the road as if expecting the dog to come trotting up it, as she did after the daily walks.

Alatna's worst morning was that of the very day when Cris was to find the dogs at the ranch and bring ours home. She came to the door as I prepared breakfast and opened her jaws in the first inch of a howl, talking to me and looking in again and again. (She did not like to enter when Cris was moving about in the cabin.)

After he left she came in and examined the couch, where Tootch usually lay. Then, as I washed dishes, there was a noise outdoors. With a sweep of her slender legs Alatna had moved the planks from over Tundra's old dig under the cabin and in three minutes had dug halfway to freedom. I stamped on the floor and got her out.

I let her take the broom, always a favorite trophy of the

wolves. When I looked out the door, she was carrying the broom by the middle, tossing her head and prancing—the invitation to the game. No furry jaw fastened on the end of the broom. I felt sad to see her. At this moment she sat down, a paw on the broom, raised her head and howled.

Then she began taking everything out of the cabin. When, for once, I became really peeved and my voice showed it, she kissed me. No hard feelings wanted.

Next she burned her nose tip on the stove. As she started silently to trot outdoors past me, I commiserated. She answered with the faintest little squeak, not of self-pity but of hurt.

When Tootch arrived home from her fruitless fling—Rip was too old to father pups—she was as hostile as ever but Alatna seemed happier. The establishment, poor thing as it was these days, was together again.

But there could be only one real comfort for Alatna and we knew what it was and that it would be a long while in coming. It would be a mate and puppies of her own, a new wolf family. It could not be hers until the following spring and her first estrus. I thought I could barely endure my life until then.

I was already corresponding in a leisurely way with zoos, seeking a tundra wolf as mate for Alatna. The San Diego zoo replied discouragingly that it would be a long search: tundra wolves did not thrive in the climate of the main United States. But it was clear from Alatna's restlessness during Tootch's absence and from her continued grieving that she needed a friendly four-footed companion now. We ourselves were not enough. We began to hunt for a dog, one big enough to stand up and play chest to chest with a wolf. It did not cross our minds to spend money for a dog.

6 *New Dogs*

Cris arrived home one June twilight with a big yellow dog beside him on the seat of the pickup. "Taffy," he introduced the dog proudly. "I named him on the way home. He was running loose at Cripple Creek and nobody knew his name. They said his folks had moved away and left him."

The dog received my greeting pleasantly but uneffusively. "He's old," I objected doubtfully, meaning too old to be playful with Alatna.

"He's not so old," insisted Cris in the teeth of the visible facts.

We fed Taffy—his soft short coat was dull from famine—and stood watching while he ate. He looked to be part Boxer, part golden German shepherd. His legs were scored with raw red bullet grazes: teen-age boys had used him as a target, so the man said who had pointed him out.

The full moon sidled from behind a skyline crag; the draw was flooded with moonlight and black shadows. We led Taffy down to the home pen to meet Alatna, penned there.

In the branch-broken moonlight the lovely wolf stole to him, whining ecstatically. He was the first big canine male she had met since Arctic died. She was taller than Taffy but she lowered her height by crouching, and reached upward to kiss his face. Stubbornly he averted it. But unsure of his status here he glanced toward us: how much of this revolting female effusiveness was he supposed to take? He stood it for fully two more minutes, then snatched and snarled at Alatna.

She shrieked. From her lips poured such a torrent of injury, rebuke, entreaty as we had never heard before. She protested screamingly this total, this insupportable rejection. She begged, reproved, pleaded. Her voice went up and down. There was nothing in it to remind one of the big howls. The performance was dazzling. It was the whole hot lava bed from which should emerge the raw material of language. Her mouth widened and closed incessantly. Now the whole line of her white teeth gladdened our eyes, then was concealed. Her tone changed as her cheeks retracted or moved forward, from reedy to plangent to pure soprano.

All this virtuosity was matched by an equal torrent of grace and gesture. Not one part of her, not for one instant, was motionless. She crouched beside Taffy, her head tilted away, a paw lifted toward him, trying to stroke his face. Diseuse, *ballerina assoluta*, she was both. Her eyes were black in the moonlight and glittering with feeling. Beyond this there could be nothing further in the way of communication short of words themselves.

At least once daily from now on we were treated to this performance and it never failed to stop us in our tracks. Taffy would stay, there was no doubt of that. He bit Alatna but he noticed her. He took her mind off her troubles to that extent. He was not a dog one could love but one had to respect him. Self-contained, friendly but not effusive, he had, we thought, been a man's dog. He loved to hunt rabbits and ground squirrels.

Tootch, on their first walk together, undertook to lead him, bringing forth all her old lead-dog bag of tricks. He could not have been more indifferent. Soon Tootch learned a new way of behaving—to go along as an equal with another dog. She was happy. Taffy was an excellent spotter of small animals—cottontails and golden-mantled ground squirrels. Too excitable and human-oriented for that herself, Tootch loved to help bark and chase. The two often cornered but rarely caught anything. There were too many hollow logs and coverts under rock ledges as hiding places for the animals.

Still Alatna needed a dog. "Bring her a *little* dog," I urged Cris, meaning a young one that she could win as a pup and play with wolfwise later on.

Cris brought home a little dog. He was a white feist, a year old and a dedicated wolf-hater. All the first evening and part of the night, to judge from the barking, he chased Alatna. Obligingly she ran. In the morning, with dreadful wolf gaiety, she prepared to kill him. When I ran down to rescue him, she retired and watched. The distraught little dog sank a tooth in my palm as I lifted him.

This time I would go myself to get Alatna a dog. Taking the white feist in an old wolf travel box, I drove down to Colorado Springs and returned the dog to the headquarters of the Humane Society, where Cris had obtained him. Two attendants led me along a cement aisle between cages of clamoring dogs. Each cage, humanely opened, to its own outdoor runway. The dogs rushed out to bark at each other in the runways.

One attendant directed my attention to a silent black pup, alone at the front of a cage, watching us with cloudy green eyes. "Part Weimaraner," said the man with respect. "He's three months old. He should grow to be a fairly big dog."

I took him. But when the man started to place him on the seat beside me in the pickup, "No," I directed firmly. "In the cage in the back." I knew wild animals better than tame ones and I thought this love-hungry puppy would try to escape.

Cris named him Swagger because he switched his rump in walking. Alatna adored him. We refrained from touching him because it was all too clear that at the touch of a hand he would forsake the wolf for human beings.

In spite of being denied human affection, Swagger was a lucky pup to have the wolf to play with him. And how she played! Every evening as the sun sank behind the ridge and the air cooled, the two raced up and down the steep bank in the yard pen until Swagger gave up and stood on the terrace halfway down the bank, barking madly. Alatna would sail up and down, leaping clean over him every time she passed.

We would stand watching, pleased to see her having fun. We shouted praise: "Look at that Alatna run!" At every shout her speed and gaiety increased. She would not lead the puppy off into the big pen: she knew we would not follow.

One fact troubled us: Alatna grew thin, the puppy fat. One evening as Cris worked late at a new project, building a guest cabin—the "Wolf Den"—over on the bench, he heard from the aspen grove along the bench an oddly familiar wolf whine. He stole over to see what was going on. Alatna stood giving the old wolf puppy-call, used by Trigger and Lady. Swagger ran to her and she gave the well-fed puppy her own supper by means of the wolf disgorge.

This is a perfectly controlled act for the feeding of the young. A hunting wolf can bring home food in its stomach and give it up at will. We need not have worried about Alatna. She discontinued this practice when ready, how we did not observe.

I was witness, however, one hot afternoon when she terminated a bad habit Swagger had of crowding between us when I knelt to pet the reclining wolf. We were in the aspen woods when the pup crowded us and this time Alatna's eyes flashed, her teeth showed, she snarled. I pushed him away. He never did it again but stood to the side, disciplined.

All the same he was still Alatna's cherished foster puppy and Tootch and I presumed dangerously with him. I spanked him lightly once, to make him stay in the big pen when I left it, and Alatna cried. That evening Tootch bit him and he cried and Alatna kissed him. She was concerned for her puppy but still docile; I did not know that someday she would "turn another corner," psychologically.

In a July gully-washer the brook rose. Where it flowed out

under the fence at the far end of the big pen it washed under the rocks and they settled, leaving an escape hole; Alatna found it and went out. She went to Cris, at work on his ruined road. He dropped his work and took her for a walk. She came home with him as far as the side gate, but there, after first trying vainly to take Swagger with her, she definitely turned away and left.

The old sad arrangements were made for the possible return of an escaped wolf: the big drive-in gate down the avenue was propped open; the small near gate to the big pen was closed, confining Swagger and Tootch. At night Swagger was shut as usual in the home pen, to keep him used to staying there in case Alatna returned.

As I went down to release him in the gray morning twilight two days later Alatna rose from under the twin spruces in the big pen and walked quietly to the small gate, waiting for it to be opened for her. A little of her spirit had died on the trip she had just taken. Wolves have a power of recognizing a hopeless situation. We felt sure that she had gone to find her fellow wolves, believing she could find them if only she could get free.

I had a custom of taking Tootch and Taffy for a daily walk. I always tried to spare Alatna's feelings, as a walk started, by sneaking the dogs out the door at the far end of the greenhouse, onto the roadhead there, and up the side of the ridge, hidden from view by the greenhouse until near the top. I would rush the dogs silently over the top and start them down the other side. On the morning after Alatna's return I started the walk early, while the draw still lay dark in the shadow of the mountain. The dogs got away from me and ran to the top of the ridge, where they paraded to and fro in the sunlight. Looking upward from the shadowy draw, the shadowy wolf saw the two bright dogs, gold Taffy and red-gold Tootch, against the blue sky, trotting past red ponderosa pine trunks. She howled.

I took a chance. Shouting the old tundra-walk call, "Let's go for a walk!" I threw open the side gate. Alatna raced out and up to the dogs, Swagger with her.

There was no need to fear that she would run away. She wanted one thing only—to be accepted into the "pack." She fawned along beside the dogs, entreating them with voice and action. They paid not the slightest heed to her. From now on she went along daily on the walks but did not again waste all her time courting the dogs. She ran off on side trips, returning often to check up on us and court the dogs briefly.

Alatna and the dogs differed in their ways of hunting and observing and relating to each other socially.

The dogs would corner a golden-mantled ground squirrel or a cottontail in one of the many hollow logs, then run from end to end of the log, barking excitedly and peering in at the squeaking, terrified animal huddled at the center. They ripped splinters from the outside of the log near the center, or tried to drag an end of the log with their teeth.

Compliantly Alatna would run to and fro with them a few times, then take a look into the log for herself. Perceiving at once that the animal within was impregnable, she was through with it; she wanted to range onward, though she would linger bored for a few minutes before drifting away. I had to compel the dogs to come along or they would have run and barked joyously and futilely for hours.

Alatna gave her attention to things the dogs ignored. Taking a new route one day, we came to a forest-service signpost at a fork in the trail. The dogs ran past it. Alatna halted. She circled the post, crouching her hindquarters warily and looking up with lucid, black-pupiled eyes at the pointing arms. For her no data were to be disregarded. The dogs limited their interests. Or perhaps the wolf was more curious and interested in things for the sake of their variety.

The day came I had long dreaded, when we met the Herefords. I had always taken precautions to avoid them by inquiring where they had been seen or by climbing the ridge before a walk to look over the arm of the pasture below. But on this day we ran squarely into the cattle. The dogs came to me when I called. But Alatna dropped to the ground, watching the Herefords. She reared her chest and neck unbelievably tall, making a perpendicular line from the ground up to her chin. Unlike dogs, wolves can alter their shapes a great deal. Alatna stuck where she was until I started away with the dogs. Then she ran to us.

She always acted concerned and responsible about Swagger. Twice he got lost. The first time he had recklessly raced after Taffy, whom he could not keep up with. Alatna, returning from one of her customary brief jaunts aside, came to me, expecting to find him with me as usual. I called him. She threw up her head and uttered a short call, then listened. Finally she acted on her own; she took off down the mountainside, apparently on his track, looking back when I begged her to stay but disregarding me.

Then Swagger appeared and I threw joy into my voice, calling her, but she did not read it. But presently she appeared in an opening in the trees below and I attracted her attention and held up heavy Swagger so she could see he was here. She came bounding up!

On another occasion Swagger disappeared and again she came expecting him to be with me. I called him anxiously and this time she stayed with me. She trotted to a nearby rock I had not noticed—always a wolf is conscious of the total terrain; it was the highest spot in the marsh where we were. She stood on it, calling too. Suddenly she started off up the wooded mountainside. Running away, I thought, but in a minute she reappeared—with Swagger.

One day he jabbed his little muzzle and mouth full of the screaming white misery of porcupine quills. Soberly I turned the walk toward home. Alatna stood beside him when he stopped and tramped with his small front paws on the quills in his nose and shook his head desperately. There was nothing she could do but she was concerned and solicitous. She hovered over him, watching him. The two big dogs ignored it all; they ran about chasing squirrels as usual.

"Why, she's socially responsible," our friend Carol had said once, watching Alatna just in the pens.

Alatna was quite capable of losing us all on a walk. Taking a new trail up the wooded mountainside one day, with the dogs I followed its switchbacks, but Alatna, sure of our direction, cut across them, going upward. Abruptly they ended and the trail made a right-angle turn. I did not call to Alatna, sure that she would soon notice our absence and come back to find us. She must have gone clear up the mountain to a cross trail; Cris found her tracks up there the next day. Finally I stopped at a spring by the trail and waited with the dogs. I called. My voice must have echoed confusingly from crags nearby. Probably Alatna could not place it even if she heard.

Starting worriedly down the trail, I heard a sound, her call, I thought, far below. I stopped. Was it just a fly zooming near me? I walked on, then heard it unmistakably, the grieving call. I answered, wildly shouting and joyful. There was nothing but sunshine and silence for so long that I thought "No contact." Then she burst out of the brush below the trail, panting and smiling, her ears laid back, her eyes shining. She "loved" us all, then found a shed deer antler and let Swagger take it. She lay playing with him on the duff in a glade among the trees.

A week later, contouring the mountainside, I lost the whole crew myself. I went downward, shouting, to get out into the clear where they could find me. Tootch was the first to appear, panting down the mountainside, then Taffy. Then I stood on a bare ridge and called "Swagger, Swagger," with a breaking voice, and "Alatna," up at the tree-hidden mountainside. I was discouraged when up from below, up the side of my ridge, Alatna panted to me, Swagger toiling after, so pleased, kissing me, kissing Taffy, kissing Tootch. I took them down to water.

Swagger rejected Alatna on the walk one day. It saddened me. He cast in his lot with the dogs. She goaded him to notice her by making him cross—nipping at him, dashing at him, until he barked and chased her furiously. But his rejection was not final: he still played with her sometimes.

About once or twice a week Alatna ran away. She would come homeward with us as far as the aspen flat at the foot of the draw. There she stopped and watched us start up the steep road, then she turned away and left, not to return until evening. I was not worried. I knew exactly where she was going and that she was welcome. She was going to visit Sox, an elderly, shy little cattle dog, at the Cox ranch downriver. Unlike gruff Taffy, Sox was civil, a policy of discretion on the part of a little dog toward a wolf four times his size.

Mr. Cox hoped for a pup from Alatna that Sox, before too old, could train as his replacement. Dubiously I mentioned the disparity in the sizes of the two animals. "He could stand on a rock," urged Cox. Still I was doubtful. "There's lots of rocks around your place," he insisted hopefully. But the decision did not have to be made for a long while yet, not until Alatna's first estrus, in the following spring.

Cox said Alatna's lie-up was the top of a wooded knoll, across the pasture from his ranch house. It was a better lie-up for a wolf than Crag cabin afforded. Hidden by trees, Alatna commanded a various view—the grazing cattle, the running Tarryall, an occasional car passing on the dusty road between the creek and the ranch house. Back of the knoll rose the mountainside and no doubt she saw wild animals from time to time. Cox, standing by the ranch house in the mornings, would call to Alatna and she would howl in answer. (But I doubted that her howling was a sign of happiness.)

Sox easily became footsore, so he rode to work in a jeep. Alatna would follow along, running through the midst of the cattle with eyes only for the little dog in the jeep ahead. "I'll make a cattle dog of her yet," said Cox with proud amusement.

On the first morning of September, waiting at the gate for the walk to start, Alatna walked among the dogs with the sinuous sidewise ripple of perfect health along her spine.

Out in the green pasture beyond the ridge she played excitingly with Swagger. Water flying, she dashed along a brook troughed deep in the grass. Leaping, shoveling up mouthfuls of water as she came, time and again she raced up and down the brook. Swagger ran after her, then, outclassed, stood barking. She leaped over him each time she came, carrying her body low and weaving it flexibly, rippling sidewise. Once, quite wolflike, she took a bad fall, hit on her chest and scooted.

Coming homeward with me and the dogs about ten o'clock, she turned away at the foot of the draw and started back along the road. I followed, calling. As I rounded the end of the ridge to where I could see the pasture, she was already on its far side, small with distance, running along the dusty, sunny road in the direction leading toward the Cox place. I called again. As my voice reached her she paused and glanced toward me, then ran gaily onward.

She did not return that evening. She was still not home the next morning, when I had to leave for a brief speaking trip. I disliked to leave with her gone. When I returned a day or two later she was still absent. At noon of the day she ran away, almost at the boundary of the Cox ranch, she had been shot. The fellow who shot her had called to her as she ran—as usual, harming nothing—and she had stopped and half turned. He had shot her through the chest. She fell but when he got over there she was gone, he could not find her.

Like Arctic's, her death would be slow, taking perhaps a couple of weeks. Every day or so I had to ask, "Would she be dead yet?" The answer was the same: "Not yet, I 'm afraid."

"Let her dream of Mr. Arctic and the others," I prayed. "Let her dream of the gaiety and gentleness, the quiet lazy talk before sleeping, the eagerness of setting out. Let her feel happiness again."

There was nothing here for her to come back to. The dogs shunned her—the pup, even, that she had done all she could dream up to win. She went on to nothing when she left us. She was the only one of her kind in the whole country.

Constantly my thoughts were with the wounded, dying wolf.

"Alatna, always by nature so light, so gay, sweet and confident. Patty paws at the gate; racing, laughing, on the wet sand in the Killik. There is no place in the world for you," I thought, addressing her silently in my mind. "It is impossible, you must feel, but that there are your kind somewhere, that you can find them. There is no place, Alatna." She had been so quick and sweet and loving when the dogs and I used to come home from our surreptitious walks to her, waking under the spruces, beaming and gentle—pleased but always rejected, always going off alone then, and crying.

One evening, two weeks after Alatna had been shot, we were at dinner, facing the black brightness of the picture window. "Would she be dead yet, do you think?" I questioned.

Gravely Cris replied, "I think she would have to be dead by now."

There was a tumult at the gate outside the window, the dogs shrieked, there was a deep moaning. She was here! I ran outdoors.

Alatna sat at the gate, leaning against it and crying.

7 *Wounded*

As I knelt bathing Alatna's wound the morning after her return and she lay on the plaza beside the rock wall, I made her a promise. Unforeseen the moment before, it came of itself, as lightly and naturally as one of the family stepping into the living room. I said silently, "Alatna, I will give you seven years of my life." The numeral, too, came of itself. The promise went to its own place in my being and I gave it no further conscious thought at the time.

We had not seen the wound the night before. I had noticed only that Alatna kept one forepaw curled up. Disregarding me and the dogs, she had made her way among us, straight toward a worthless half skull in the shadow; she must have spotted it as she entered the gate. With it in her mouth, not taking the easy way around by the path but her accustomed route down over the terrace wall, she gathered herself and jumped, to go to her old lie-up under the spruces.

Hurrying indoors to prepare food for her, I said, bewildered, "She didn't even flick down her ears for me."

Laconically Cris replied, "She's in pain."

The next morning we had seen that her left foreleg dangled from her shattered shoulder; there were white splinters of bone in the raw flesh. She must have turned as the bullet sped, so that it had missed her lungs.

Cris drove off to find someone to advise him what we could do for her. I did the only thing I knew to do—bathed the festering wound with warm salt water. She accepted the bathing as if it were natural and proper, like another wolf licking her wound. I bathed the wound daily from then on.

The one priceless thing I could have done for her I did not think of. I did not notice that she kept her forepaw curled at all times; even in sleep she did not uncurl it. I could so easily and gladly have opened and stroked the no-doubt aching ankle.

The veterinarian Cris consulted said it would be useless to look at her, that a wild animal in its terror at a stranger might do itself worse hurt than the original wound.

"The worst problem a wild animal has when it's wounded is food, and she solved that when she got herself back to you. Feed her well. Give her all the calcium you can choke down her. The wound won't start to heal until the last of the shattered bone works out and that may take four months."

It had taken Alatna two weeks to retrace the miles run in two hours, going, and during that time she could have had water but no food. She had eaten ravenously the evening she arrived. But from then on she ate listlessly. The one food we knew she would have liked we could not obtain—whole carcasses to pick and choose from, with all the organs and connective tissues in them. Not even a whole rabbit could be located. The wolves had liked rabbits almost better than caribou. The bleached remnants of rabbit, cellophane-wrapped, from a supermarket in Colorado Springs she picked at listlessly.

My problem was to find something in which she would accept the white calcium powder. At last I recalled a certain brand of breakfast sausage that, in more affluent days, we had often bought. I tried her with a small bowlful of this sausage mixed with the powder. She ate it eagerly. Thereafter we bought it by the case and she never failed to eat her bowlful of it.

For her lie-up during convalescence she chose our bunk in

the bunk nook at the rear of the cabin. We moved out and slept in the unfinished bedroom in the near end of the greenhouse building. I even left the bedding on the bunk, merely spreading a sheet over it, in case it added a trace to her comfort.

She required that the outside door be left open: she was not going to be "trapped" in a closed cabin. If she heard the door being stealthily pushed to, she came hobbling from the nook to escape. Even when the hinges were oiled she detected the movement of the door and came out. When later the temperature fell to twenty below zero, the door still stood open and, plates on laps, we ate our meals huddled in parkas beside the little cookstove, its oven doors open for added warmth.

She knew the sound of our pickup but a strange car driving up put her into a panic. She came out from the bunk nook so precipitately that often she slipped and fell, until a strip of black building paper was stapled from the bunk to the door; this gave her claws purchase and she did not fall again.

Once there was a crash from the bunk nook and Alatna came hobbling out, her eyes dazed-looking. She had rolled off the bunk. This did not happen again.

One sunny afternoon as I sat silent, sewing in the shadowy main cabin, and the sun shone into the bunk nook, I heard a faint moan, not of pain but of utter weariness of pain and confinement. Alatna must have wakened and thought herself alone. That was the only sound I heard her utter in complaint at her wound.

We thought the wound was her only problem but to her there existed another, finding a mate for her coming estrus, due in the following spring. It was not too soon for her to become "engaged." One day the gate was left unbolted and Alatna actually went away, injured as she was. Discovering her absence soon enough so that I was sure I could overtake her and sure, also, that she would be heading to see little Sox, I set out after her, taking the dogs to allure her home.

In the marshy main pasture the dogs went curiously to examine bushes at the side. Positive she would have gone farther than this, I did not look but she must have been there. Her body was not up to her spirit. Soon after I came home she came too.

Her resilient spirits rose. Often she went out and hobbled around the pens. We were looking forward with pleasure to an event we had by now arranged and that, we were sure, would give her a big lift—the arrival of a wolf we had purchased to be her mate. He was to come from a zoo in La Crosse, Wisconsin; the purchase had been arranged by correspondence. He was not a tundra wolf; we had not been able to find one of that kind. He was a timber wolf, hence would be smaller than she, but still a wolf, to play with her and be her mate.

We received word that he would be shipped from La Crosse by train on the afternoon of October 17. Two days later, in the afternoon, Cris drove down to meet the train.

It was twilight when he returned. I ran out to meet him and the new wolf. Cris was walking so dejectedly toward the cabin that in spite of a glance past him toward the pickup that held a strange cage, I inquired with anxiety, "Didn't he come?"

"Oh, yes," Cris replied listlessly, "he came all right. . . . He's only a pup. A card on the cage says he's seven months old."

This was a blow. Alatna would not have a wolf mate for her first estrus; there was no time to go through the process of hunting again. Still, I thought, some degree of pleasure was possible for her: she would have a wolf to play wolfwise with. In my mind also was the thought of the little zoo wolf's happiness in the wild, relatively big terrain here.

We walked to the pickup. In an open-wire cage, exposed to view on every side, cowered a slight, motionless gray figure. He looked small to us, but was probably about full grown and the zoo attendants might have assumed that he was sexually mature as well. Cris had already named him Coonie, because of a black, raccoonlike mask across his eyes.

We lifted Coonie's cage from the pickup by the handle bars on the sides and carried it down to the home pen. We set it beside the walrus rock, facing the big twin spruces, black in the twilight. Cris raised the slide. The wolf did not move.

"Coonie," I said softly.

He did not move. I hesitated: he could turn his head and bite. Then I put my fingers through the wire and gently touched his harsh-haired side. This was one of the only two times I ever touched Coonie. He still did not move nor turn his head.

We blundered. We should have gone away and left the little wolf in his opened cage, the only known object remaining in his whole world, to venture out when his courage sufficed, even if it took him all night. But I could not wait for the happiness of the two wolves to begin. We tilted the rear end of the cage, obliging Coonie to step out onto the ground.

Again he stood motionless, looking straight ahead. Beside him rose the gray, nubbled granite of the walrus rock; before him loomed the unfamiliar black trees. The cool, darkening air was filled with strange scents—sage, pine, unknown animals.

Alatna, in the big pen, was passing beyond the spruces on her way toward the brook and the path to the bench. I called to her but she went on. I seized a handful of wet yellow straw from the floor of the cage and ran to bring her. Already at the brook, she glanced back as I called. I knelt and held out the straw. She came back to see.

She snatched the straw from me and flung it. Her eyes at that moment I shall never forget. They glittered black with mad question, undarable hope: was earth, then, the scene of resurrection and paradise restored? For one moment she hung, her eyes on mine. The next moment she flashed past me, running toward the home pen. I ran after her.

She went up to the wolf. He cowered slightly but did not move otherwise—not even his eyes, fixed straight ahead. He was not one of her own. He was a stranger and a pup. Alatna turned away indifferently and left him.

Coonie gave us the measure of what we had done for our wolves. They had been strong and agile. Coonie was feeble. His belly sagged. He could get himself up onto the table rock only by hooking his toes over the edge of boulders ahead and dragging himself upward. The gate was left open by day but he was too timid to leave the home pen. He beat a trot path along the fence in the most concealed corner of the pen, the dark corner back of the table rock.

The disruption of the little wolf's inner world must have been inconceivable. I don't know which had traumatized him worse, the manner of his transportation or his early youth caged. As to travel, our wolves had been exposed to its nerve-racking horrors seven times as long as Coonie and had emerged unshaken in confidence. But they had been shipped in the travel boxes, closed on all sides but one, thus affording them concealment. As to Coonie's puppyhood, he had lived all his life in a tiny pen with his mother and litter mates until the minute he was removed for shipment to us. Both his parents and grandparents had been born in captivity.

I learned about Coonie's youth when I wrote to the zoo inquiring what his diet had been, even to brand names if any. He was refusing to eat. I duplicated the diet as best I could and gradually he began to eat. He gained strength and at last could bound up over the boulders onto the table rock. But he still clung to the home pen.

He was standing on the table rock one night when I made a second blunder with him. It was late when I heard him howl—the first sound he had uttered since his arrival. He stood on the table rock, his head reared toward the stars. His howl sounded infinitely sad. It died away on the highest note, as if the sound had gone on across a thousand miles of space, toward the only fur and friendliness he knew. Softly I answered, to let him know there was friendliness here.

So! Even in the darkness and stillness he had not been alone. He never uttered another vocal sound that I heard.

8 *First Estrus*

ONE MORNING in early November, as I was preparing breakfast, Alatna snatched the hot water bottle at a moment when I could not leave the stove. When I finally followed her I yelled, hoping to keep her on the move so she would not stop and chew on the rubber. Over on the bench at the head of the path, she laid down the bottle and came back to meet me, dipping her head sidewise, her ears sweetly flattened, lifting her paw, her eyes black and happy. A grand commotion had been had, a grand success!

But later in the morning she could not get anyone to play with her. The dogs lay on the couch. She appeared at the open door, bringing the most impressive bait she could muster in her meager kingdom, the half-head of a cow. Her face had a mingled expression: she looked as if she felt she had something, she hoped for a game and a challenge, she felt gentle and gay and mischievous. Her ears were up and half pinned back. She looked demure and pleased and hopeful. The dogs lay still.

In the afternoon the sky turned dull and gray. Alatna, lying on the terrace, looked upward with indifferent eyes. I could have "licked her wound" then—I still bathed it daily and she still received the bath like a licking from a friendly wolf—but I did not realize how unhappy she was. She came in and courted Tootch and Swagger for a minute, then went outdoors.

Soon from the far end of the bench came her mournful crying, pitiful, like a grieving woman's voice. It was not the great reverberating mourning howl, somber and almost awe-inspiring in this rock chamber of a glen, but a faint, grieving cry.

Regardless of her feelings I went for a walk with the dogs. Alatna was no longer taken along. Our return was joyful. Alatna wagged her tail and danced up and down on her one good forefoot.

Then it snowed hard. Snow filled the air. And Alatna on the plaza in front of the open door was rearing upward, leaping, very tall and white-bellied and deep-furred, into the thickly falling snow. Her eyes black, she was looking upward into it and clopping her jaws on it at the top of her leaps. It was a strangely beautiful thing to see her joy and participation in the storm—the beauty of her body going up into the snowflakes, the beauty of her eyes watching and of her mind, that she was interested.

In the morning she was gone; the gate had been left unbolted again.

Four days passed, then one afternoon the Cox men, father and son, jeeped over to let us know that Alatna was staying around their place. They thought her hangout was up a lonely wooded gulch, a mile from the ranch house, where entrails had been placed after butchering.

We went at once, unbelieving, hopeless. In the dark, where the road trace failed, we left the pickup and walked on by flashlight.

We howled near the entrails in the gulch. Unbelievably in the night came the deep answer. Then Alatna's eyes appeared in the beam of the flashlight. It had taken perhaps ten or fifteen minutes for us to join each other after the contact howl. We gave no food. We hoped she would become hungry enough to come home of her own accord. She acted pleased to see us, but extremely wary and quailing.

We skipped a day, then went again at evening. This time we had brought along a travel box and choke and leash chains. We set the box on the ground at the rear of the pickup. The chains I placed in the big pocket of my padded jacket.

She answered our howl and came, this time from a different direction. Nervous but willing, she followed me toward the travel box. But at the slight chink of the chains as I lifted them she was gone. I leveled the flashlight to my forehead and with the beam swept the bare space between me and the nearest woods. It was simply not possible that she could have got to them so quickly.

Two days later, going at twilight, we noticed a shallow depression leading away from near where I had stood. Like a flash she must have gone to it and crouched away along it. A wolf's nerves must vibrate ceaselessly, like a seismograph, recording data about the terrain. This "recording" is something different from the open, active wolf curiosity.

We continued to make rendezvous with Alatna every other day. Going early once, before the sun was down, we saw something that disturbed us. Her answering howl sounded faintly, from far away. It took half an hour for us to approach each other. She came out of the woods on a low bluff and stood at its edge, looking around for us. What disturbed us was that her white-looking fur shone like a star in the low sunlight: she would make a good target for rabbit hunters, who would swarm the woods on the coming holiday weekend. We could not wait any longer; she must be got home.

We left the Cox ranch house at seven the next morning, leading Sox on leash as bait. The day was sunny. We heard Alatna's deep howl to the northwest, answering our first howl. We climbed the hill in that direction. She did not answer again. We searched vainly among rocks and trees. Discouraged, we thought she must have lain down already for her siesta, but we howled one more time. This time she answered. still on to the northwest, across the Tarryall. We crossed by a foot bridge.

Alatna came running toward us over the morning-shadowed

pasture, from the knolls at the foot of the mountainside, overjoyed to find us, and especially Sox, away out here. Showering attentions on patient serious little Sox, she followed along happily as we started northwestward, on the long trailless walk toward home.

Once we came to a trail going up the mountainside and followed it, hoping it would turn in our direction. It turned the other way and we had to backtrack, concerned about the extra labor for the crippled wolf. At any moment she might give up following us. Next we ran into canyon walls and again had to backtrack to climb above them. Down by the Tarryall afterward Alatna lay down on the dry leaves in a bare aspen grove and so did we and Sox. We urged her on after a while.

We stayed along the mountainside, avoiding the few houses along the road. Once Alatna found a shot rabbit. I never saw a rabbit eaten so fast. I did hope she felt we had something to do with getting it for her. We feared the meal weakened our hold on her. Soon afterward she lay down once more and slept for an hour and a half, then we urged her onward again.

The dogs had been left shut in the home pen and the drive-in gate left standing open at the end of the avenue. Would Alatna go as far as the gate and then at the last minute refuse to enter it? Our tension increased. Also it worried us that Sox was tired and becoming irritable at Alatna's unceasing attentions. Nearing home, about three o'clock, I thought wryly, "The next five minutes will tell whether our day's work has been in vain."

As we reached the big gate the dogs up in the home pen began to shriek and Alatna sailed in and flew hobbling up the avenue straight toward their fence. She must have felt that the big welcome was for her. I thought of a shipwrecked captain's wife who, with other survivors, after days of rowing in an open boat, had come to a town at evening. Gala lighting was strung along the bluff—to welcome them, she supposed. It was for a party at the governor's mansion. "I had never felt we were to be pitied until now."

The footsore little bait dog had to walk another two miles

with Cris, to our nearest and only neighbor's place, for a ride home.

How joyfully we took up all the inconveniences and responsibility—yes, pain too—for the joy of having the young wolf with us again and safe from being shot. The door was open, the room cold at breakfast so that she could come in and out for bites of food from her fork. I dismantled the cabin at night so that she could sleep in the bunk nook undestructively. Pillows went into the closet, ornaments to the cupboard; the typewriter was put away. We were glad to have Alatna home.

Yet I knew that for her infinite boredom and loneliness lay ahead. But I was happy, on the morning after her return, to see her eyes calm and contented. Yet she would long for the fluent wolf tails flying, the friendly wolf faces, the tireless invention and romp, and fur stretched acceptingly by fur to sleep. She would long for travel and adventure, for finding smells and trails, and just for being at liberty and going—to see what was up by that rock or mountainside. It would hurt us to see her stand on the table rock, looking with those keen eyes of hers for movement that was not here—searching this empty glen with her wolf gravity and competence.

I felt that she had never thought of her life except as basically free. We were trying the impossible once more: to keep a wild-born, half wild-reared wolf captive yet feeling free at heart. We took for granted that we would not fail.

In mid-December began a month of drastic changes.

Coonie changed. For the first time I saw him give a sign of playfulness. On top of the table rock, as Alatna and Swagger romped and leaped below, then whirled out of his sight, he bounced playfully, half participatingly, to the other side of the rock to see them come into view. Later he rolled, actually luxuriating, on an ice glare in the pen. Maybe now, I thought hopefully, his spirits would pick up. I liked the little wolf and hoped he would become happy. But when Swagger made a playful jump at him, he cowered away, his spine humped, his tail between his legs, like an old beaten man.

On this same day Alatna played bewitchingly with Taffy. She held her ears erect and pinned so tightly together that they touched. They slanted backward slightly, in a way that prolonged the line of her profile. She bent her neck sidewise with an infinitely graceful toss and tilt of her slender head. She lifted a forepaw. The expression in her eyes was indefinable, so alive, so challenging. Winsome was the word for her. How winsome a wolf can be!

I did not know that for the first time I had witnessed the mating-play position of Alatna's ears. I supposed that her estrus was still three months distant.

At the close of December, two and a half months after his arrival here, Coonie again did something he had not done before. He entered the cabin one morning while I was still in bed. Scared, he urinated a puddle when Tootch bustled in. I think the open door had looked to him like the entrance to a hiding place.

On that same day he entered the big pen for the first time. He did not fear gates as our wolves had done at first. He was near Alatna in the avenue when I came home with the dogs from their walk. Always I hoped for increasing friendship between those two. After letting him enjoy his liberty until twilight, we walked him without difficulty through the gate of the big pen and into the home pen. The next day he presented himself in the yard at the closed gate to the big pen. I opened it and he went in. Again toward twilight we walked him to the home pen by staying behind him and separated. We started at the far end of the bench and did not see him nor each other until, at the gate, we glimpsed the slight gray form stealing through, to slip along beside the home-pen fence and in at its gate.

Daily from now on Coonie enjoyed the big pen. The little zoo wolf utilized its delights by trotting through them to the fence at the far end of the bench and beating a trot path there. In storms he sheltered on the dry duff under a big fir tree near his trot path.

In January occurred a major change for Alatna: the last splinter of bone worked out of her wound and it healed rapidly. Now, with grief, I saw how I could really have helped her. She began uncurling her paw and putting it to the ground. The ankle was ankylosed. Gradually the skilled, fingered paw would splay, to accommodate to the rigidity of the ankle. If I had massaged the ankle no crippling would have showed. Strangers saw none now, nor did I when she ran, but she limped a little when she walked.

Her moods went up and down. She put me away with her nose one day when she ran up the bank to the group of dogs and to me, kneeling on the deck. I waited with hands and smiles to receive her, but she was coming to the dogs. She gave me a little push with her nose. After that commanding touch at me, she smiled and licked—or rather flapped her willing tongue out at—the dogs' growling faces.

She got Coonie out of the home pen and played with him. She played with Swagger. He "attacked" her, growling like mad and wagging his tail at the same time. His little mouth was only big enough to grab less than the length of her deep fur. The only place where he could get hold of anything substantial was her paws, of which the safest selection was a hind one. Sometimes he actually tumbled her over.

Taffy would have to be her mate for her approaching first estrus, since Coonie was too young to mate. Taffy was a different dog from the one that had come to us the preceding June. His yellow coat shone. In the fall it had been a delight to see the golden dog running beneath the golden aspens. For a long time after he came to us he had eaten absorbedly and uncritically, having been hungry too much to question his food. After a long time he would look over at Alatna's pan of delicacies but revert to his own pan. Nowadays he was masterful and confident. I looked him over as sire for Alatna's pups. His shape was good: he was not a bred monstrosity as many dogs are; he had a good basic feral shape. His thick muzzle was unworthy of the wolf's slender one, but on the whole Taffy would do.

Alatna changed toward Tootch one morning. Screams from Tootch called me outdoors to see her and Alatna fighting in the sunny snow down at the foot of the bank. I ran down to rescue her. She was fighting hard but was outmatched. As I joined the fight Alatna still aimed her bites unerringly toward Tootch, but the dog in her desperation inadvertently sank a tooth into the inside of my wrist, miraculously missing tendons and artery. I got her out of the pen. She could not enter the pens again but became what we called an "outside dog." She had asked for her trouble: for twenty-two months she had shown implacable hostility toward patiently courting Alatna.

I was kneeling beside Tootch one day near the side gate, attaching her leash, when I became aware of Alatna's tall form beside me. It seemed so natural to have her near that for a moment I did not grasp that she had escaped from the yard and that the dog was in danger. The next moment she seized Tootch by the side of the neck and began shaking her.

I threw open the gate and by the leash, so luckily attached, dragged the heavy, fighting mass into the yard pen and shut the gate. For a few minutes Tootch had to take it, live or die, while I preserved the captivity of the wolf. One glance showed me that it was not by the yard pen she had escaped. I ran down and closed the gate into the big pen, then I turned my attention to saving the dog's life.

From the cabin I grabbed whatever was near and threw it. Alatna would withdraw but not far enough for me to get Tootch into the cabin. I grabbed the mop and hurled it. It clattered and broke. The noise drove Alatna far enough down the bank so that I could haul Tootch into the cabin, shoving Alatna off with the heavy door just as her teeth started to sink into the dog's flank.

Tootch's neck swelled to the size of a watermelon but she soon healed. The help of vets was still far from my life.

Alatna became so low in mood that to cheer her we let her kill a big white rabbit in the home pen. We had discovered boys near Lake George who raised rabbits. We placed the squashy, cage-stultified creature beside the fence and went out, closing the gate. Alatna withdrew warily and observed the rabbit. It made a feeble bewildered hop or two, then with one pounce Alatna killed it.

I thought we would not do this again. But she spent the night in the home pen for the first time in months, though the gate was open. She did not even come up to greet the dogs in the morning; did not, in fact, leave the home pen until Cris went down and praised her for her important accomplishment. Then she brought the skin up into the yard. She had practically skinned the rabbit as she ate it.

We let her have one more, but this time she was so indifferent that thereafter we killed the rabbits first and gave her the warm bodies. She seemed as well satisfied.

In mid-January, prematurely, I thought, Alatna entered her first estrus. Cris carried in a roll of wire one day to mend the fence and left the side gate open. Alatna went out. A heavy snowstorm came on, snowing us in for the first time.

On the morning of clearing, when the sky was pure and each puff of air sent white powder towering into the blue from firs or crags and no one would want to be indoors in such magic, Cris decided to try to make it out through the drifts in the pastures to the public road, which would no doubt be plowed. He wanted not to hunt Alatna but to go to Fairplay, the county seat. When he had not returned an hour later I knew he must have made it.

In the afternoon I was astonished to see a strange dark jeep floundering up our road. It was the Cox men, with word of Alatna. They were astonished to learn that Cris had gone out. No trace of his passage remained in the pastures. The men had backed their jeep and rammed the drifts or shoveled through them.

They thought Alatna's estrus must be about over, for she was becoming cross with Sox; they had shut him in the barn when they left home. Alatna was staying in the woods nearby. Taking a travel box, choke and leash chains and sleeping pills, I drove off with the Coxes. In spite of drifts even on the public road, they took me around by way of Lake George, where I bought hamburger in which to give the pills.

At their ranch the men went to the house in order to give me a free field. Standing in the snow near the barn, under a sky now overcast, "Alatna," I called softly and doubtfully. To my surprise her gray form came almost at once from the dark fir woods nearby. She crouched around me, so pleased that for a minute I thought I could leash her, but no. She was too wary, as always when free. I held out the clumsy pills, one at a time, in balls of raw hamburger and she gulped them eagerly. The Coxes had considerately refrained from feeding her.

She stayed near me until she felt the drug taking hold; then she went to an unused corral and lay down on the snow. Presently I moved toward her quietly. But she got to her feet and went off into the woods. Staying at a little distance, not to panic her into flight, I followed. She dropped down in a small opening in the woods and I stood beside trees, waiting for the drug to take deeper effect and feeling keenly my ignorance of what I was doing to her. I could only guess at the number of pills needed for anesthesia. Sharp at my heart was the fear that I might have gone too near the margin of her life.

Finally I stole toward her with the chains noiseless in my hand. She got up and staggered away. I did not put her to the run but went back and waited by the trees. She fell down.

At dark Cris joined me silently. He had followed our tracks into the woods. He took the chains and stepped toward the prostrate form but Alatna got up and he saw that it would take an all-out run to catch her. Doubtful whether he could overtake her, he came back and together we stood waiting.

Her body lay black on the dark snow. The temperature was twenty below zero. I feared the effect of the drug would soon be lessening. By the dim snowlight we went to the ranch house.

Mr. Cox brought his lariat indoors and suppled and warmed it through his hands, then coiled it and hung it on one shoulder. With the only flashlight among us, his torch, he led out into the night. Single file we followed, with his son.

Alatna lay as we had left her. As our small black procession headed by the torch neared her, she got to her feet and ran, with definitely better control now. We ran after her easily, not to panic her. Beside a log she stopped and faced us.

"Now," commanded Cox quietly. He handed Cris the torch and lifted the coil from his shoulder.

The rope flew straight. The noose settled true and perfectly over Alatna's head, tightened around her neck, caught a twig from the log and failed to tighten enough. She pulled her head out and ran, Cris at full speed after her with the torch. If we lost her now, she might drop down in an exposed place and freeze, sleeping off the drug. Cox fell out because of a bad heart. Young Cox and I ran after the disappearing yellow beam among the black trees.

The penumbra of light ahead vanished. We held our direction and ran on. I ran out into a level-looking strip of snow that proved to be a gully, sank in but scrambled up the other side. In the black woods again I hit a single low strand of barbed wire and pitched headlong to the snow with a gashed leg, jumped up and ran on.

"Hurry!" came Cris's voice from ahead, with a note of desperation in it.

We ran in the direction of the voice. Again it came, more desperately, "Hurry!"

There was a patch of yellow high in the blackness at my left. I started straight upward toward it, unable to choose a way, pulling myself up over rocks. I heard young Cox coming close behind.

Silhouetted black against the yellow light from the torch he was slowly swinging in his hand, Cris stood at the mouth of a low rock cavern, facing Alatna, who was crouching to and fro along the back of the cavern, her eyes, brilliant with terror, fixed on the beam, seeking a chance to slip out past it. From Cris's hand held out silently as I passed him, I took the chains. Speaking Alatna's name softly I knelt beside her. She did not glance toward me nor take her eyes off the light as the choke chain went easily over her head and enclosed her neck. I handed the leash to Cris and we started down the hill.

Young Cox waited just below the level of the cavern, having given us all possible chance for the capture. He gripped the leash along with Cris. Cris alone could never have held it. Alatna strained powerfully away. Not once did she turn her head toward the hands exposed on the chain.

The travel box sat on the snow at the rear of our pickup. Alatna was crammed into it so ruthlessly that I gasped. One of her hind legs was doubled back into an unnatural position and, dazed by drugs and terror, she could not protect herself and gather it to her.

The next morning her first estrus ended in a flood of blood. She lay at the foot of the granite tower on the near end of the bench. Taffy stayed near her for an hour or so, then left her. She lay alone over there for two days.

Another year must pass before Alatna could again have a chance at puppies and a family.

9 *Tootch's Puppies*

WHAT helped Alatna through this year of waiting was Tootch's puppies, her first litter. There were five of them, fathered not by gruff Taffy but by gentle, solicitous little Swagger. Each one was different. At first I thought of them as mere tools to help Alatna in her loneliness but I soon learned to cherish them for themselves. No animal is merely a tool.

As Tootch gave birth, hidden in a nook on the back plaza, Alatna stood at the fence, fascinated by the sounds and smells of birth. As soon as the puppies grew to be toddlers and came into view she watched them with equal fascination.

Not realizing it, I dealt harshly with Tootch: I made her share her puppies with hated Alatna. Not sure of what might happen, I took them into the yard pen one day for Alatna to meet them. Tootch, outside the fence, watched with bitter eyes as Alatna nosed the babies and presently made the wolf disgorge for them.

After this I put the puppies into the pen for a while each day. Eventually they would become "pen dogs," remaining in the pens or cabin all the while. First, though, Tootch had some happy times with them. I took them along on the daily walks and Tootch ran agilely ahead through the spring snow while the puppies floundered determinedly after her. She seemed to enjoy being a leader again.

One April evening when the puppies were two months old Cris met me on my return from a short absence with a mystifying story. Clearing the yard pen, he had started to bear out one of the puppies by the scruff of the neck. The puppy shrieked; it was too old and heavy for that treatment. Alatna had been far away, down under the big spruces, when Cris started. In one flash she was behind him and gave him a nip that rent the seat of his levis and made a pink mark ending in a blood-thick scratch.

This was not the Alatna I knew. My Alatna was the one revealed in an incident a day or two later. A commotion on the bench sent me flying over there just in time to see Coonie making off through the snowy woods toward his far trot place with a puppy hanging by the neck from his jaws. He gave me a scared look, dropped the puppy and fled onward. The puppy ran to the fence and slipped through its wide mesh into the forest outside, where she stood, crying shoulder-deep in the snow, reddening it with blood.

I knelt by the fence and held out my hands through it, coaxing her to come. It would have taken too long to go around to her outside the fences. She came and I started working her, all legs, through the wire as gently as I could, aware of my risk. Alatna stood beside me, her heart in her ears and eyes. The puppy screamed as the wire touched her bitten neck, and Alatna jumped her big forepaws up onto my shoulder and bit my cheek very lightly—no mark—in her excitement and will to prevent that cry of pain. Then she licked my face a brief nervous flip or two. She had not meant to hurt me and the kiss was to redress the balance. This was "my" Alatna.

And what had Coonie thought he was doing? I think he was kidnaping a puppy to take off and have to himself. By now the puppies were big enough to ford the brook and they often played at the foot of the sheer face of the pen tower, while Alatna lay near. Coonie would steal from his far trot place and

The wolf Alatna.

A wolf-dog pup caresses Alatna's mouth—a gesture used both to show affection and to ask for a handout. Here it is simply affection.

Alatna and adult wolf-dogs—Kuskokwim behind Alatna, Kotzebue and dark Katmai.

After opening the screen door—a skill she taught herself—Alatna holds it open.

Whose ham for breakfast? It was our pleasure not to control or discipline Alatna overtly.

Alatna solicitously nuzzles the loser in a dog-puppy fight.

The junior choir.

The young female wolf-dog Katmai playfully crouches in ambush behind a tree.

Alatna "waltzes" with her dog mate Baranof—an activity she always required during mating play.

Alatna kisses Lois. (The dog Swagger looks on.) Only a few days before giving birth, Alatna carries her pregnancy inconspicuously.

Alatna and a wolf-dog pup lap at a pool in the brook.

Alatna and Mr. Shy after she accepted him as a mate.

Friendly tug of war over a worthless beef tail (Alatna *left*). The animals did not fight over food.

Torn between desire to come in, and distrust of a person in the cabin, Alatna takes an observation leap at the door.

Alatna and her coterie, dogs (*left*) and wolf-dogs.

Alatna yawns happily on the deck table. As usual one pane is missing from the paw-and-nose-marked door.

A favourite lie-up near the woods in the big pen.

Alatna liked the bunk.

Siesta (Alatna in foreground).

Alatna and puppies.

A wolf-dog puppy escapes through the wide-meshed fence.

Katmai and Tippie.

Alatna visits welcoming dog pups at the entrance of the den.

A howl (Alatna is in the centre).

Alatna and pals in the lie-up place.

Alatna and guests in the den.

Young wolves during the freeze-up time. The wolves on the right are shouldering each other, a common wolf play action.

stand at the edge of the woods watching the puppies. If he was like any other wolf we had known, he would have felt an impulse to feed and fondle them. Perhaps on this day a venturesome puppy had rambled near him and he had grabbed her, biting down harder than intended in his agitation as I rushed toward him.

Alatna exercised all her social responsibility on behalf of the puppies. A cry of pain from one of them and she sprinted up to do something about it. As I started to leave the big pen one day she ran down to the closed gate to meet me. When I got into the yard pen, she was so obviously pleased that I was flattered —for a moment. She looked so sweet, her eyes radiant, her ears back. She was full of emotion. She leaped up to put her front paws on my shoulders, kissing me and moaning with emotion. Then she told me something in a brief wow, still with radiant eyes.

I seized what it was all about. A pup was barking trouble, desperation and entreaty. As I hurried up to the plaza, Alatna bounded ahead. She said "Ow!" and even threw her head back and howled briefly. The pup was in trouble and I was coming to help. I would straighten it all out. The pup had got itself outside the fence near Alatna's lie-up beside the rock wall and had scrambled around the foot of the wall and up onto the hillside. It was shrieking because it had no idea how to get back in. I brought it in and Alatna pranced around it, touching it searchingly with her nose. No injuries on it, it was all right to a wolf's nose.

Meanwhile Tootch had sat outside, where she was at liberty to go to the pup and nose and lead it to familiar ground, but she had sat with perfect indifference to the pup, her eyes only for me. The deprived ex-lead dog craved human approbation.

If Alatna could not help a distressed puppy, she sought a scapegoat on whom to take out her feelings. One morning as I was dressing in the greenhouse-bedroom, a sudden commotion in the yard sent me sailing barefoot over snow and ice into the pen. A female puppy was struggling up the bank, surrounded by all the other pups, two of which were biting her. She was shrieking. Alatna, of course, was frantic. When the pups reached the deck, they knocked the little female down, biting her.

"Open the door," I called to Cris. Alatna was nervously tearing into Swagger and I took the instant of her distraction to lift the chief attacker off the badgered puppy, which scrambled up and ran into the cabin just as Alatna reached and mouthed me. "That's all right, Alatna," I cooed, shutting the door which separated the fighting puppies. I was still sure of my Alatna.

Then one morning I met another Alatna. The two strongest, most spirited pups began a battle. Equally matched, neither would give up; the fight went on and on. It was the first such fight there had been. When I tried to separate the pups, Alatna wrenched the stick from my hand and assailed the zinc washtub I thought to use in frightening the puppies apart. Instead, I had to hold it between her and me. At last when she and the fighters were down by the gate to the big pen and I thought of somehow getting it closed between them, she flew at me, face to face (a punishing, not an attacking position), and I was lucky to shut myself into the big pen.

She had hovered over the puppies, crying and touching them with her tongue. As they screeched, her frenzy had mounted toward any scapegoat she could find, since she could not do anything to stop the fight. For once I had been the scapegoat.

After the exhausted puppies separated, Alatna licked each wet-mouthed fighter.

From now on fights were numerous and they always drove Alatna wild. She never joined them, she fumbled to stop them. She and her siblings had not fought except on two offbeat occasions. There had been squabbles but no fights. The wolves were peaceable among themselves.

At last one day Alatna hit on a novel way to break up a fight. A strong puppy had jumped on the weakest, smallest puppy, Gray Lady, who shrieked. Alatna first put a paw over the aggressor, then took her gently by the tail and hauled her away upside down. Then she kissed her but nosed her belly admonishingly and also growled. After that she stepped over and kissed Gray Lady, who still lay in a heap against the gate.

Tail-hauling was an old wolf custom but used for other purposes than stopping fights. Trigger had used it to "rescue" a puppy from me. He had used it to detain a dog absconding with his meat. The wolves had used tail-hauling in play. Now Alatna had been inspired to adapt it to breaking up a fight. From this time on she tail-hauled the aggressor away. She never erred about who was the aggressor unless she was too far away when the fight began. Then, arriving on the run, she made her selection of the probable aggressor.

This meant that danger could hit quickly, out of the blue. I was going down the path one day with Alatna and three of the big pups. Alatna, divining my goal, the big pen, ran on ahead. Suddenly there was a cry of pain. A pup crouched doubled up and yelping, a hind paw caught in the new ground wire laid along the base of the fence. Without thinking I stooped to loosen the paw. The puppy snapped at me and cried again. At this I realized my danger from Alatna. Smoothly I rose and walked away, barely in time. She arrived on the run and, growling, nabbed one of the pups. The one caught by the wire managed to free himself.

Alatna used all her wolfish social arts to win and hold the puppies. Resourcefully she tried to attract them to herself if they were engaged with someone else. She would come into the cabin, look on the table or couch for something to carry out, and hurry outdoors with it—a pillow, a plate, even a paper sack—and the puppies flocked to see.

One day when Swagger had them engaged, playing with a toy, Alatna found a superior attraction. She went to an old dig of hers, an earthen tunnel, idle since she had been shot, and enlarged it. The pups watched inside the tunnel, helped dig or lay on the rim watching.

In this first year of her sexual maturity, Alatna's courage rose in all directions, though always accompanied by wolf wariness. For the first time she ventured to drink at the dogs' water pan under the sink, though with many a wary quailing glance up at the "ceiling" she was risking her head under. She stretched out fast asleep on the bunk while I washed dishes quietly a few feet away, a deeper relaxation than she had permitted herself even when wounded. I never moved a chair nor lifted one while she was in the cabin, for to do so would be my last resort to scare her in an emergency.

Her fearlessness was a difficulty when for some reason I wanted her to get out of the cabin. I could get her out only by stampeding outdoors myself, as if frightened by something without. Even this ruse was failing: I had cried "wolf" too often. But usually as I waited in the yard, puffing excitedly to keep up my pretense, she would lag to the screen door, push it open and saunter out, fairly certain there was no cause for alarm but obliged by wolf caution to make sure.

She had mastered the screen door of herself. It had taken her a year of persistence and three wrecked handle fittings, but now she controlled the screen as easily as we did. Coming indoors she had to do three things almost at once: grasp the horizontal handle bar in her teeth and pull it toward her, in the reverse direction from that in which she wished to move; hook her strong curling toes around the edge of the screen and pull, not push it; and finally, with a foreleg swung backward like an arm, hold the screen open while she entered.

That summer rural electrification moved up the Tarryall valley and we had a line run to Crag cabin. Alatna took note of each step of the installations. She absented herself while the electrician was at work but as soon as she heard his car going off down the draw she came indoors and observed each new object. On the day a switch was installed beside the door she quailed as if hit as she started to enter, and turned her nose directly to the switch. The overhead lights, both in the cabin and under the wide gable on the deck, were duly noted. When a shade was added to the ceiling light, Alatna crouched nervously to look upward at that. The dogs ignored all.

As the puppies neared full size Alatna still came into the cabin for a trophy with which to induce them to follow her—a towel, a bag of shelled pecans, a candle. When I had everything put away I would watch her scan the room. She would jump onto the table and survey the counter at each end of it. She jumped onto the couch and reared up to inspect the high window ledge above it. I knew with amusement that in spite of my care she would discover something to take out to the puppies.

In so many ways they were seeing Alatna through this hard year of waiting. They had become her pack. We dared not discipline the indomitable five. They made a constant ripple of trouble.

Sometimes old friends drove far to visit us; sleeping space was scarce and I would sleep in the Wolf Den cabin on the bench. When I emerged in the morning the dogs jumped to kiss me. I kept a switch by the door to take out with me, to fend off the loving mob. But I let it fall innocently the minute Alatna appeared, coming around the base of the pen tower, notified by the dogs that I was up and could be greeted. She hated a raised stick though none had ever been raised against her.

That fall Swagger became an "outside dog" like Tootch. He had a falling out with one of Alatna's five and she sided with her pack dog. Swagger met her in the cabin after this but could never re-enter the pens. This was our good fortune; he was our dog at last. He was a virtually infallible watchdog. Trained and disciplined by Alatna, he was pleasant to have around. He reveled in our affection, given freely now. If I turned involuntarily as I worked in the kitchen corner and he lay on the couch, I learned that it would be to meet his eyes fixed on me with shining devotion.

Coonie was in trouble and I could not help him. The five young dogs persecuted him. His chance had been as good as Alatna's to keep the upper hand over them from their babyhood onward. But Coonie was a wolf bowed in spirit: he could not, literally could not, dominate. Waking in the Wolf Den, I would hear barking approach and would lean on an elbow to look out the back window as Coonie stood at bay on a ledge of the fence tower. His final refuge would be the end of the table rock, down in the home pen. There he would face the barking, venturing dogs and clop his teeth audibly at them, a sign not of audacity but of desperation.

It grieved me to see the little wolf in trouble and not be able to help him. I liked Coonie. I liked to please him and win his shy friendliness by small attentions when he was alone. I would kneel and hold out my hand and speak softly to him or toss him tidbits. He dared not come to me but I knew he liked my attentions.

There was only one way to help him and it was unacceptable —to pen him permanently in the home pen. He loved his poor liberty in the big pen, and his far trot place. More important, we looked forward to his mating with Alatna in her coming second estrus. I watched for signs of more than casual friendliness on her part. Penning him away from her would alienate her.

In December Alatna's mood rose as it always did to meet the first big snowstorm of the year and perhaps, too, her approaching estrus. In the falling snow she leaped straight upward, tall, and whirled completely around at the top of her leap. Her action was beautiful because of the flowing beauty of her winter coat and the happy brilliance of her black eyes.

Shortly before Christmas we declared a rare holiday: we would go together to Colorado Springs for a full day. Usually one of us remained at home to safeguard the animals.

While we breakfasted and dressed on the morning of our holiday I let Coonie into the big pen for an hour of liberty. Then, in town clothes, we went to shepherd him back to the home pen. He was quicksilver. Time and again he slipped unseen between us and back to his trot place. We would come in view of the big-pen gate, and not seeing him had to return to the trot place to comb again through the woods and along the hillside.

Finally, we drove away and left him, though it troubled me, looking back, to see that among the barking dogs at the fence

Taffy stood silent, angry, I knew, at being left behind, and especially at being left penned with the "inferior" animals.

On our return that evening I went to the yard, intending to go indoors to start a fire and dinner for animals and people. But in the yard pen the animals met me with bewildering emotion. Alatna stood up and pinned me against the logs of the cabin, a paw on each shoulder, and talked passionately to me through kisses, "M-l-r-oom." Even self-contained old Taffy came up whimpering eagerly, his eyes lucid and bright and innocent like a puppy's eyes.

I warmed the animals' supper and set their filled bowls on the plaza. Holding Coonie's bowl, I called him. No shy gray form appeared, stealing through the big-pen gate and gliding close along the fence toward the home-pen gate. Bowl and flashlight in hand, I went to look for him.

I was going methodically along the back fence on the bench, calling, when my voice failed in my throat: I did not want to call any more. Silently I went back to the cabin.

In the morning Cris came in as I prepared breakfast. "I found him," he said in a low voice. "He doesn't look like Coonie. The dogs killed him."

I went as directed, down to the woods by the brook. I knelt by the mouthed gray body. Gently I touched its side. It was hard. This was the second time I ever touched Coonie.

What a hell of dog-yells this lonely draw must have been while the killing took place. The dogs had fierce dreams for a week. None of the dogs had a mark, except one, who had a single puncture on his nose. Coonie's skin had not been broken so far as I saw.

Less than a month later, Alatna entered estrus.

10 *Baranof*

We stood despondently on the deck under the gable light one evening in late January, watching Taffy and Alatna at their courting play on the plaza. Alatna's estrus must be about over and it was clear that there would be no puppies: Taffy was a big dog but he was not tall enough to mate with the long-legged young wolf.

On impulse, unaware that the device was a common one among breeders, I went to the pair. Kneeling I elevated Taffy by his hind legs; his forelegs gripped Alatna. Forgetting their preoccupation, the astonished animals gazed around at me over their shoulders. It would never work. I went back and stood beside Cris.

Abruptly he spoke. "I know where there's a big dog. A really big dog. He's a MacKenzie River pack dog and quarter wolf." In the morning Cris drove off to Colorado Springs to try to borrow this dog.

It was twilight when the pickup rolled up onto the back plaza. In the rear of it stood the lordliest, biggest, most imposing Husky I had ever seen.

"Baranof," Cris introduced him.

Overawed, I advanced and held out my hand. The dog growled perfunctorily. Intimidated, I backed off.

We made all ready. The pen dogs were shut into the big pen, except for one male, Brownie, who belonged to Alatna as a dog may belong to a human being; he would not leave his "owner." (But she tolerated no advances from him.) He and Alatna had the yard and home pens to themselves. It was dark; the gable light shed a dim yellow illumination over the plaza. Baranof was admitted to the pens.

Alatna was enchanted with the stately dog. Like Trigger and Lady, she looked up to size, maturity, "presence." Baranof disregarded her. He applied himself to examining and re-marking every scent spot in both pens. Alatna crouched wooing beside him.

At last she retired and lay on the edge of the plaza, following him with bitter eyes. She had never withdrawn like this from Taffy, but he had never ignored her. When Baranof was ready to give her some attention, she chivied him with dangerous gaiety. Keeping Alatna's body between himself and the big stranger, Brownie barked ferociously. Baranof was in trouble.

We called him. That superbly trained dog obeyed. Flying up the bank toward us, panting and harried, he looked all wolf.

The next morning he was returned to his home. Alatna's second estrus passed in vain. It seemed happy enough. Taffy was Alatna's dog at last. Gallantly the old fellow unlimbered his stiff joints to stand up and "waltz" with the young wolf, as she required.

Her third estrus must not fail. That summer, in exchange for the promise of choice of litter, Baranof became our dog. He must become accustomed to the place and animals, and they to him long ahead of time. At first we kept him as an outside dog; later he was introduced to Alatna gradually, always in a pen cleared of other dogs.

He rode daily to work with Cris, who was mining near Lake George at this time. Baranof loved the rides. Toward his many human fans he was casual but friendly. Leaning his big forepaws in friendly fashion against the chest of one admirer, the star-route mailman, often encountered at Lake George, he twice knocked the man flat, a result that only increased the man's admiration for him and added respect to it.

Baranof was not a young dog—his former master never would divulge his age—but he was powerful and agile. His stride was so long that without seeming to extend it he could easily overhaul a fleeing rabbit. He sprang with ease from the ground to the high-tilted rear end of the dump truck, clearing its tailgate.

He came near springing to his death one day. The pen dogs had long since matured and one of them came into estrus. To reach her Baranof made a mighty leap from the hillside above the yard pen, clear across the new overhang wire and down into the pen. In a flash the pen dogs were on him. He went down under them. Worst of all, Alatna joined the attack. She would dart in, bite his hip and retire. It was clear that Baranof would be killed.

Standing beside me outside the fence, Cris said quietly, "I can get him out if you can keep Alatna off me."

"I will." I knew I would be bitten. We entered the pen.

Cris went toward the pile of dogs near the rock wall. I went toward Alatna, on the opposite side of the plaza. She retired down the bank. I stood waiting. She came up over the edge of the plaza, carrying herself low, her eyes black and malignant.

"Don't bite me, Alatna," I said in a low voice.

She sank her teeth into my thigh. Eclipsing the pain was a novel sensation: to look down and see a wolf hanging to my thigh gave me a feeling of anomaly, as if the order of nature had been overturned. Alatna retired. From behind me came Cris's voice, "I've got him. You can come out now." Unwisely I turned my back and started to walk out of the pen. She bit me again, in the rear. Then I was safely outside.

I was afraid to re-enter the pen, but after a couple of hours I had to resume my work in the cabin. Alatna had lain all this time at her lie-up beside the rock wall. She watched me. As unprovocatively as possible I moved close along the plank table on the deck and went in the door.

When feeding time came I felt afraid of Alatna but I took her bowl of supper to her. She still lay in the same place. She watched me with cold black noncommittal eyes as I walked easily toward her. In as casual a voice as I could command, I spoke to her and set the bowl at her head. Then I walked back to the cabin.

My greatest fear was that I could never feel warmth toward her again. But after a day or two I did. This incident was the nadir of our relationship. Alatna and I were not born loving and trusting each other; we had to learn. From this time on our mutual trust increased and never suffered another recession.

That fall was the one before the year of Alatna's first puppies and the publication of my first book.

In September, to my sorrow, Tootch met accidental death. In October three good dogs were killed to make the pen safe for Baranof when he should enter. One was Taffy. For him I did not grieve, though I felt sadness. His life here had been good. He was old and, outranked by Baranof, he was no longer happy. He was too wise to fight the big dog but he and Swagger had fought for the first time, I thought because their nerves were on edge over Baranof. Taffy had won but was permanently injured.

As for me, having hit upon the way to tell my story, I was racing to complete the manuscript by the deadline set, December 31.

Meanwhile our savings had run out. The mine was not bringing in much take-home pay. We dined on beans and breakfasted on the best of the animals' "blood eggs," bought by the case from the commission house in the Springs where we bought meat for the animals. We never dared run behind on our bill there more than a hundred dollars or we never could have caught up. Preparing breakfasts for all, I lined up bowls along the work counter and broke each egg separately, distributing them according to quality. Now and then one exploded; it was hard to scour the smell off the bowl. What kept our health abounding was a big daily salad from the greenhouse. The animals, not fat, were zestful like ourselves.

For me, alone at home, noon was a restless time. Everything in the refrigerator, cupboards or deep freeze was earmarked for other meals. My hungriness would be too trifling to mention, in a world hourly more filled with hungry people, except for the surprise it was to a middle-class, educated woman. I was surprised to the point of amusement, a levity unforgivable if I had not been sure this period soon must end.

A friend who went hunting on the western slope of the Rockies gave us two elk steaks the size of blankets, a windfall for us. Usually we preferred not to eat wild meat, though people dining us prided themselves on serving it, apparently on the assumption that if we were interested in wild animals we must want to eat them. But now we valued these elk steaks. Lying on top of other supplies in my wooden carrier, the first steak was brought from the refrigerator in the storeroom, part of it to be prepared for dinner. As I entered the cabin, an emergency at the stove caused me to set the carrier on the counter by the door and run to the stove.

When I looked around, Alatna was standing up at the counter, the steak already in her jaws. It was useless to try to take wild meat from a wolf but I tried. I ran uphill, downhill, around rocks and trees and towers, begging and pleading, while gay Alatna galloped easily ahead of me with the steak. I went back to the cabin and called the outside dogs and ran noisily down the road with them, in full view of watching Alatna. It would hurt her feelings. I wanted to hurt them. I felt bad.

Having gone to the big pen for some reason on my return, as I was leaving it I heard Alatna coming up the avenue behind me on the run, her paws thumping. Flinching inwardly–I had no idea what she intended to do–I stalked ahead. She hit me between the shoulder blades. I stood still but without turning my head. She reached her head over my shoulder and kissed my cheek!

Alatna stayed in the pens by the grace of God and because I was part of the fences. They needed repair but we couldn't afford the time off to mend them. About every half hour I toured them, watching for signs of an escape dig. A tour took about ten minutes. Should I set my plate of hot scrambled eggs and cup of coffee on the table and note silence from the pens, I left the food and ran to make the tour. Noise was reassuring. Silence could mean an escape dig starting. If I forgot my duty and became absorbed in my writing, silence brought me to myself with a start and I would run to make the rounds. Not until siesta time did I dare to relax my vigilance.

In the first week of January, in the year of the puppies and the book, Alatna began her first serious den dig and I completed my manuscript except for minor revising. Quickly I wrote an article I thought had a chance of selling. I sent it to the *Reader's Digest.* The reply was, Get it printed somewhere and we'll see. I rushed it off to a small regional magazine. It was accepted at once: it would be reserved for the anniversary issue, the next fall; payment on publication. The next fall! My discomfiture was only partially relieved by a good laugh.

But by fall it would not matter. Already the tide was turning and more take-home money was coming in. But it would be a long while before we could relish a bean or an egg, or I catch up on my craving for ice cream.

11 *Pregnancy*

IN THIS SPRING of her fulfillment, looked forward to for so long, Alatna revealed by her behavior that she had wolf-action patterns, different from a dog's, that were being denied. Her first frustration was caused by Baranof himself. It was in connection with her first den dig, on the bench in the big pen under the ledges of the fence tower. She was very hopeful of luring Baranof away from the plaza into the big pen, we thought in order to interest him in the den. He was totally indifferent to her because Gray Lady, one of the pen dogs, was in estrus. Alatna would go off crying.

"It's not a light thing," I thought, "that this beautiful animal grieves, no creature to meet her nature. She wants Baranof to come to her den digs and share. Share the eagerness and zest and work. Share also the gaiety and courting play."

While Gray Lady's season lasted, Alatna was so hostile to her that we thought we saved the dog's life by having her stay indoors at night. Alatna might have killed her. We did not prevent the two from meeting at times, for if we had done that Alatna would have been implacable forever. Gray Lady was a diplomat. When Alatna's hair foamed up along her spine, her eyes got brilliant and she growled frighteningly, Gray Lady wagged her tail, attempted to kiss Alatna's nose and at worst lay in a dump of passivity until Alatna's storm should blow over. Alatna would "kneel" on her—that is, crouch and extend her forelegs over the dog's prostrate body, resting her weight on the dog, pinning her down.

Perhaps it was because Baranof ignored Alatna that she was driven to the action I witnessed one Sunday morning. Going down the avenue on a routine fence tour, I stopped. Ahead, where the brook flowed out under the fence, Alatna was working hard at an escape dig. Every minute or two she paused, panting, and gazed out at the woods where she would be in, I estimated, about ten minutes. She knew exactly what she was doing.

She spied me. Her expression altered; she looked at me dubiously. I stepped forward praising her warmly for her wonderful achievement. Her face relaxed. She ran up and kissed me, then flew radiantly back to work. I sneaked away rapidly to call Cris.

She made a revealing mistake another morning, connected not with her reproductive needs but her social ones. It was a dazzling cold dawn. The pickup would not start and I went out to help push it. As it started moving down the road, Cris climbed in. I followed down around the bends of the road, pushing whenever the pickup faltered, until it reached the steep part of the road and rolled from view. I listened until I heard the engine start, then walked slowly back up the road, breathing deeply of the cold pure air and looking at the ponderosa pine green in sunlight beside the road and the firs down in the draw, still black in the mountain shadow.

To my astonishment Alatna met me at the gate and threw herself into my arms as if I had been gone a day. She measured her greetings by the length of my absence. This one must have been emotionally long. Probably she thought I had left for the day.

After Alatna entered estrus, Baranof still ignored her, riding off daily to the mine. What kept Alatna occupied was her usual varied social life with the pen dogs. Their number had been increased by puppies, older and younger, descendants of Tootch and Swagger.

The dogs liked Alatna: she not only ruled them but also did pleasing things for them. In ruling, she enforced wolf "oughts" so far as she could. These included imperatives about food. Apparently it had important social aspects to the wolf. She hated greed. She liked to bestow food. She liked her authority over it to be acknowledged. She liked to choose the recipients of it.

I gave her a carton with cottage cheese in it one day. She took it outdoors and laid it down and walked away, her tail elevated in good spirits, as the pups thronged the carton. But Goldilocks, a dog with a sumptuous golden fleece and golden eyes, was big and hoggish. Alatna bit her. She shrieked defensively. Then all remembered their manners. Goldilocks fled when Alatna turned away. (The dogs dared not flee when Alatna was watching them.) Alatna looked after her but did not follow her.

Then a pup took the carton and prepared to leave with it. As he did so, Alatna sauntered by and the pup made the most graceful of obeisances: all in one move he curled one forepaw under his chest and bowed his head. Alatna past, the gesture of submissiveness so instantly assumed was instantly banished; he got hold of the carton and made off. Some baby pups had withdrawn a couple of feet from the carton as Alatna passed and had remained unmoving.

But though the puppies bowed to her moods, they ran joyously after her when she had a trophy, and ran up impromptu to stand up on their hind legs and kiss her muzzle. She would lay her ears back, pleased, but often she growled. This growl meant nothing and they knew it.

Once when Gray Lady hogged a dripping pan, driving off the pups, Alatna pinned her down. Another time she jumped on hoggish Goldilocks, who grabbed a box of dog candy Alatna intended for the pups.

Alatna had obtained this box of dog candy by her own efforts. She opened the screen door, came in and partly closed the big wooden door in order to get at the low cupboard back of it. Then with her teeth she expertly opened the latch of the cupboard door—another of the skills she had taught herself—and with her claws hooked open the door. Inspecting the contents of the cupboard, she selected the unopened box of "candy" to take outdoors and give away. Her series of actions, including the attack on poor Goldilocks, had culminated in what I thought was a non sequitur: she nosed my posterior. A nudge like this was never followed by a bite; it was merely an assertion of authority, a warning.

Alatna hated dog fights as much as ever and still jumped on the aggressor, usually Goldilocks.

She performed substitute actions, redirected activities. I would not let her enter the cabin once and, foamed up all along her spine, she jumped on first one and then another of the dogs thronging around the door; she had them all in an uproar of screeches and escapes.

One February day she grabbed *Last Tales* by Dinesen—just arrived—and was out with it. I yelped, controlled my voice to sweet jest and followed. No success. Her eyes were black. She jumped at the crowding dogs around her trophy, laid down by the travel box in the yard. She jumped at me. But I stood my ground, gauging if I dared, dogged to get my book back. And what did she do? Jumped up to my face, her paws on my breast, and kissed me—a substitute action, I was sure. A sign of relenting was that she jumped on a dog and did not return to her book. I got it back, minus a few pages.

Alatna had a special technique for keeping the dogs under control when they threatened to get out of hand. They gave her at least two big chases a day and she enjoyed them, up to a point. I looked out once and watched a chase just starting. An amiable young female dog was barking at Alatna. Alatna stood, tail slightly elevated at the rump. The other young dogs ran up and ringed her, barking. All at once her impulse to run hit the boiling point and away she went, pups ahead of her, pups behind her. Over the rocks, around the rocks they ran, some dogs running to cut her off, a subsidiary chase starting between two dogs. Alatna whirled, stood at bay, then gladly ran again. It did my heart good to see her having fun.

But the dogs had no bounds. She was a different kind from theirs. They went too far. I had gone back to my work but a pup squawked and I looked out again, to see which one was getting the technique. The littlest, shyest pup lay paws in air. Alatna made growling snatches at it, big chewing motions that did practically no damage, while the pup averted trouble by screeching. Alatna turned away. When the pup was sure it was ignored, up it scrambled and circled to her head, wagging its tail like mad. At least she made life eventful for the pups.

And this was her favored technique whenever the dogs were about to get out of control: she pounced on the least and humblest one of them and made it screech. This put the fear of God into all the rest. They retired and watched solemnly. Coonie could not, simply could not, have done any of these authoritative, resolute or generous acts that Alatna performed. He had been a wolf cowed in spirit.

Alatna reached the peak of her estrus in early March and Baranof stayed home from the mine, devoting himself to her at last. While the height of estrus lasted, going down to her bower under the big spruces was a unique experience. She "voluptuated," she offered herself to caresses. It was like going to the boudoir of a loved beauty. (The lie-up under the spruces had been roofed over and bedded with hay.)

Even when I took food to her she might offer it a smell but before eating she gestured in a way that mystified me the first time it happened: she bowed her neck sidewise, tilted her head aside and looked up at me with black, gentle, gay eyes. At the least touch of caress on neck or shoulder, down she lay in the full "wolf greeting," asking to be loved. Only then would she eat, and she did so as if eating were a continuation of the pleasure. Her actions were not those of specific sex, but of sex generalized, flirtatious. She "voluptuated."

We realized now that her second estrus need not have been in vain: this year she had not "stood" for Baranof until the seventh of March. The year before, Baranof had been introduced not too late but far too early. Her estrus lasted seven weeks. I wondered whether this prolonged time might give wild wolves a better chance of finding mates.

Mid-April marked a change. Baranof had Alatna shrieking for the first time since the height of her estrus. He had stayed with her, sacrificing his daily rides, longer than most dogs would have done. Perhaps his quarter-wolf heritage affected him. But a wolf wants more than even the best of dogs is prepared to grant. She wants to be "loved," wants to play, court a bit, even though her heat is past. But now Baranof rode off to work daily again, and Alatna had one more frustration to endure. In a few days she had adjusted herself. Life was back to normal, except for the overwhelming fact that now Alatna was pregnant.

When I entered the yard in the mornings, always the wolf on the far side of the sea of dogs looked to my eyes. A glance allayed jealousy—eyes met and that alone, without the touch of hands, reassured her: she was number one "dog."

Coming indoors one morning to build a fire, start breakfast and the day, I let the horde of seven dogs and Alatna surge indoors with me. I always flinched as I opened a door or gate to let all through. Had I overlooked some risk? Was I letting together two that would fight? So as usual I took a hasty glance around the cabin as I entered. Was there anything—pillow, book, our breakfast ham, that Alatna could seize for a trophy? All clear. I took off the stove lids and crumpled paper to start the fire.

"Oh, *Lois!*" Cris's deep reproach sounded from the side yard. Alatna was standing on the table, looking out the wide-open window, with nothing between her and freedom but wolf caution.

With one soft step I was at the table, my hands around two

bony hind legs. Alatna looked around at me. I reached past her to the window and wound it shut.

The screen was absent from the side panel of the picture window nearest the work counter because I had insisted on a way of entering or leaving the cabin in emergency without going through the yard pen. A regulation had been imposed about this unscreened window. No one, ourselves or visitors, was ever to leave the cabin unoccupied without first winding the screenless window open just far enough for a hand to be inserted from without, to wind it clear open.

Another regulation, for Alatna's security, not ours, concerned the side gate. This gate was to be bolted by anyone entering, even though another person was in the back courtyard and walking toward the gate. Some afterthought could cause that person to turn aside to the storeroom or go back to the bedroom. Certain safety precautions had been taken about the gate itself. A backstop had been nailed along the sill so that, if it did happen to be left unbolted, it could not be pushed open by the wolf but would have to be claw-hooked open. Also the bolt had been set on a slant, to defeat opening by teeth.

Alatna, as always, demanded to enter the cabin at will. But nowadays we were less ready to eat in a cold room than we had been when her loneliness was greater. If refused admission, she did one of two things: either she went off to the far woods and cried or she tried to break into the cabin by her own efforts. She would stand up at the door and drum with both forepaws together on one of the lower panes of glass. She thumped so hard that sometimes shattered glass flew to the back of the room. A supply of spare panes was kept on hand, but we had given up on the lower lefthand pane, the one over the door handle and nearest the end of the deck table. That remained unglazed, a wolf hole, through which Alatna's slender head could look, to inspect the cabin and communicate with us inside.

We were eating one day when Alatna wanted in and, refused, started drumming. She had broken so much glass lately that Cris scolded her, I think for the first time.

"I'll spank you!" he assured her.

She looked at him, withdrew to the travel box in the yard and looked back. Some thought had struck into her.

"Don't scold her," I said.

Her act and look reminded me of a recent occurrence. Up on the ridge, on a dog walk, Goldilocks had jumped on her two sisters. I scolded her, forgetting that Alatna could hear me. Goldilocks started homeward and I followed, to let her in the gate. I dared not open it. Alatna stood on the other side of it, black-eyed and furious at me. I did not want to go away and leave her in this frame of mind toward me. I knelt and "explained" my action earnestly to her, condoning it. She looked steadily into my eyes. The anger left hers; they remained cold and black, but with some inexplicable expression. She walked away. Something had gone counter to her free will.

On these two occasions we had forced some unwelcome awareness on her, some hint, perhaps, of the terrible power of our kind. We had touched her in her sense of power-to-cope. The tundra girl met the planet master.

As Alatna neared term we worried about her den. She had three digs going and we did not know which if any she would use. She had dug a back tunnel to the den under the fence tower, her first dig, but she could not join it to the front tunnel. Dynamite could not be used to help her out: it might wreck the tower. Her second dig ran back under a big low rock near the table rock. The third dig was the tunnel she and her fellow wolves had dug long before, under the front end of the table rock.

We decided to offer her a better den. We would give up the Wolf Den cabin temporarily for her. An entrance was sawed in one side and a passage from it to the bed was walled and roofed. The bed was walled too, making a den of the space under it. I don't think Alatna ever put her head into this cold dark spot.

She was doing a prodigious amount of work on the old dig

under the table rock, using her crippled forepaw as well as the other. It was ten feet from the eave of the table rock to where the tunnel met solid rock. She excavated a chamber there, with a low granite roof slanting to the ground. A middle chamber widened the tunnel also. Between the two chambers intervened an obstacle rock, not large but one that Alatna had not been able to remove. To get from the middle chamber into the back one she had to go up around this rock to the left.

It looked as if this was to be the den. I cut dry, sweet-smelling wild grasses on the hillside and crawled in with them and made a soft bed in the back chamber. Alatna did not rest until she had scratched out every blade.

She was near term but carried her pregnancy inconspicuously, as if the weight were distributed along her furry belly. A stranger would have said merely that she was in good flesh, full-bodied.

One day in early May Baranof had his second and last fight with the pen dogs. This time Alatna did not join the fight. He was able to handle the rest. Alatna stayed over at the far side of the plaza; participation through intensity of watching was in her every gesture. Afterward she jumped on anybody around, Brownie especially. I entered the cabin by the free window. No doubt I could have walked in but better to be safe.

"She likes Baranof," commented Cris.

She had not attacked him but neither had she helped him. I wondered whether this was the way she had acted while Coonie was being killed.

In the final days of pregnancy Alatna was clearly under emotional stress. She showed it in various ways.

On May 5 she killed one of Gray Lady's puppies, one of a litter of four that I had never bothered to introduce to her. They lived on the back plaza, the "courtyard," where they had been born. Through the fence Alatna had taken an ambivalent interest in them. Heedlessly, on this morning, I set down on the plaza the hot, nourishing-smelling bowls of scrambled eggs for the pen animals before feeding the hungry little puppies in the back courtyard. The doomed puppy squeezed through a crack between logs extending from the front of the cabin near the less-used side gate. We had not thought even a tiny puppy could get through it.

"Here, little puppy," I called from the gate—I dared not open it—to orient the little crying thing. Alatna lay on it, held a paw heavy on it; she was excited, eyes black, ears very alert.

Entering by the other side gate, I held her bowl of scrambled eggs between her nose and the puppy. I even took a chair outdoors, to frighten her off with it. Her meaning look toward it, her reach of the shoulders toward it, though not leaving her prey, not rising even, were deadly. I took the chair back indoors. When her jaws went over the puppy and I saw she would kill it I went away from the sounds and sights.

First, though, while the puppy lived, Gray Lady got into the pen and went toward it. She came away. She did not dare the wolf.

For two hours, not coming to eat breakfast, Alatna lay under the big spruces, oriented toward the little soft heap that lay as if sleeping, out in the open. Later she carried it to the bench and left it. She did not eat it. (Weeks later I found the skin only, back of the fence tower.)

Once more Gray Lady got into the pens. I was thoughtless and let that happen. Deaf to my calls, to offers of food, she went to the spruces where Alatna lay. She cowered, she wagged her tail but doggedly she went to the wolf. Alatna made one pounce on her. I stood silent. Gray Lady lay silent. For once, danger was beyond placating. When she dared move she followed the scent where the puppy had been carried about. She left the pen of her own accord. Alatna menaced her.

I wondered what effect my nonparticipation, and worse, toward her achievement would have on Alatna's feeling and future action toward me. Wolves are very "proud" of accomplishment. (But even in the final, stressful days of pregnancy she was to be affectionate toward me.)

It was an immense relief when Cris came home that night and I could tell him the story. He listened thoughtfully. "Funny," he said, "how people can't help attributing the whole complex to an animal, if they get hold of one part or point. If wolves are gentle with us, then wolves must be dogs. And if they are undoglike at any point, then they are 'bad.'" The serious response eased my sense of horror.

The following day Alatna was cold toward two dogs that "belonged" to her. They were Bobby and Wallie, sisters from a litter of Gray Lady's. Wallie was especially devoted to Alatna and wild toward human beings; I had never been able to touch her.

Alatna bit on Bobby until the dog screeched. Bobby stayed completely out of her way all the rest of that day. Wallie was different. Alatna made for her too, head lowered, ears forward, eyes cold. "A kill," I thought. But Wallie did not retreat; she stayed near Alatna, who lay at the gate to the home pen and her den. Wallie did not sleep but watched. Once Alatna rose and started toward her, stalking as for a rush. I called and she stopped.

Later Wallie lay stretched out beside Alatna's dig under the low rock. Safe, evidently. Alatna was now lying on the balcony under the eave of the table rock, at the mouth of her den-tunnel. (She did not stay in the den for more than a few minutes at a time.)

On this same day of her coldness, Alatna began "den-caching." I gave her a special treat, half a beef heart. She cached it near her den. The dogs dug it up and ate it. Nevertheless, Alatna cached devotedly, though vainly, for the next two days. Evidently she wanted meat near the den. She may have sensed her coming helplessness. I thought that in the wild a wolf mate might have helped her with the den-caching. Perhaps obscurely Alatna felt the need now for a real mate.

Alatna denned on the eighth of May. First, though, she greeted us on the plaza at the top of the path. She ate breakfast. About ten in the morning she went into the den.

This was a day of wild mists split by thunder, crashing among high crags above us. Snow fell. It hailed. The brook was a stream jumping white among green on the mountainside back of the glen. Flood was feared out on the Tarryall.

Alatna had no supper. She did not come out of her den and I felt doubtful about crawling to her with food. It snowed again that night.

The next morning when I got up, about six, Cris said he had looked in at the mouth of the den and she had laid back her ears for him. He had praised her lavishly for her wonderful den. So, emboldened by his experience, I crawled to her along the tunnel, taking three bowls—diluted canned milk, eggs, water—and setting them down in turn ahead of me. Alatna was so thirsty she almost choked as she drank the milk. She gobbled the eggs, refused the water.

I was puzzled, when I backed out crawling, to hear her cry and see her head sticking up at one side of the obstacle rock. What did she want? I had shut Bobby and Wallie with the other dogs into the big pen. Guessing that she wanted the dogs, I opened the gates, but big Goldilocks and others mobbed her. She had emerged onto the low balcony but she retreated into the den. It seemed that I had done the wrong thing.

Two hours later I crawled in with more milk and another egg, and was a bit surprised that timid Bobby went along the tunnel ahead of me. I had to let her eat with Alatna. I was considerably more surprised when the unnoticed shadow at Alatna's back lifted its head and was Wallie, laid warmly along the wolf's back in the tight V of the rock roof and the earthen floor. Bobby followed me out and again Alatna cried. "She wants loving company," I thought.

From this point on she was restlessly in and out of the den. Twice, on the balcony, she looked around questingly, as if seeking something or someone; looked away out of the pen—not at the clustering dogs outside the closed gate but at the empty hillside. For the wolf mate that should have been with her, to kiss and encourage her?

Out at the gate, to the dogs, she uttered a strong crying like the wolf puppy-call. It was not a pain cry, it was social, as if she sought the reassurance of companionship. I let all the dogs in, but, carrying herself lightly, she hurried back into the den.

Out once more she clopped her jaws audibly, catching saliva, and panted heavily. She went to the gate again but this time I did not let the dogs in. This was a gray day of melting snow, chilly but not cold, after the snowstorm in the night. I felt worried when right beside me she took a position as of excreting; her hind quarters trembled. If she gave birth to a puppy here, in the mud, how would I get it back into the den? But she hurried into the den of her own accord.

Every few seconds she clopped her teeth together. She uttered a hoarse cry every few minutes. I worried about the food I had given her; had I done the wrong thing?

Bobby and Wallie slept, each curled outside the gate. They had been with Alatna or near her since she had denned, the day before.

By noon I was extremely worried. Her labor seemed too long and hard. She was no longer youthful; three estrus periods had passed. If something had gone wrong we could not help her now. She uttered short cries about every two minutes, sometimes not ending abruptly in mid-air but falling in a descending "oo." Mostly the sound was a blare, ending in an upward "oo," broken off. A second or two after each came a hoarse grunt. The cries became brief harsh monotones.

Once more she came clear out to the gate, crying not the pain cry but the begging-company cry to the dogs gathered outside. She looked distraught; one hip was black with mud. She returned to the den and I came away. It was 12:25.

Cris, about to leave, met me at the door and said, "She wants you with her." I looked. Sure enough, she was standing on the balcony, looking up toward the cabin. I smiled serenely at Cris without a word; none would pass the immovable block of my determination not to go back to the pen. There was no reason for it that I could see: it was just there, peaceful and immovable; I was not going back. Cris went on away and I heard the pickup leaving down the hill.

I was sitting at the table a few minutes later, writing, when I noticed that all was silent from the den and realized that the silence might mean a puppy. I went to the pen.

Alatna was curled in the den, her back toward me, licking something, then her rump, then the other place, and there was a small reedy cry. The small fresh reedy cry of Alatna's first-born.

I stood listening and grinning. The May sun warmed softly through the cloud.

12 *Alatna's First Puppies*

WHAT does a mother wolf do when a woman crawls into her den, later conducts a food war over her puppies, when dogs lie among the puppies and the puppies themselves defy her? Does she become hysterical? Bite the woman, chase the dogs, punish the pups? All these troubles and more befell Alatna in the season of her first puppies. Yet I had been so sure for so long that puppies of her own would put her in paradise that I saw her as happy regardless of the evidence. In reality she was often anxious and frustrated as she tried vainly to safeguard her puppies in accordance with her wild behavior patterns.

I myself caused her first anxiety over the puppies. Returning from the dog walk on the day of the birth, I heard one last pain cry from the den. Soon afterward I took in a pan of water. Alatna reached toward it but rejected it. The point of interest was the expression of her eyes. Brilliant, light-colored not dark, they looked terribly apprehensive, as if she were ready for defense. I felt a bit worried, so I praised her for her wonderful puppies, which I dared not touch, and withdrew.

A soft May evening came on, mildly sunny, mildly warm. The storm was over, the sun was out. I took milk to Alatna. She drank thirstily and licked the pan and this time she was relaxed. She lay with her puppies nursing. Our Alatna nursing! After all these years. Yet it seemed normal, as if things had been this way forever. By now I had seen many dog mothers on this first day, when mother and pups are as one circulating being, so absorbed, so needed by one another.

From now on I fed Alatna regularly in the den, crawling to her with rabbit and eggnog. What I saw by flashlight when I reached the middle chamber was strangely impressive—the bright-eyed wolf lying with her puppies in the rear den under the low slant of nubbled granite. The puppies, of which there were four, were the soft neutral dun color of wild wolf puppies.

The dogs barked before we were up one morning, letting us know Alatna had come out of her den. She came out again that afternoon, when there was a happy commotion on the plaza as I fed the dogs milk after their walk. She was a different Alatna from the one that had denned. Thin, unkempt, she seemed a mere bundle of emotions. Feverishly she kissed the dogs, kissed me. She glanced toward the pans of milk; I thought she would go to them but evidently she had not time. She hurried back to the home pen, followed by the dogs. Her actions there made them understand that they were to come no farther than just inside the new inner gate. (The left front corner of the pen had been fenced off, forming a triangular vestibule or catch-pen, with two gates, the old and a new.) I did not realize how seriously she feared the dogs on behalf of her puppies.

Before long Alatna felt up to flirting. She flirted with Cris in her old playful way. Pitifully she flirted with Baranof, who ignored her totally. Starting down the path to the pen, she curled around looking back at him, her head clear back to her hip. She returned and tried to run alongside him. Obviously she yearned for him to go to the den and was trying to lure him there.

Cris caused anxiety for her one evening. He took her a highway-killed jackrabbit. She came halfway from her den to the inner gate to receive it. He started to pass her, going toward the den. "I only meant to peek and try to see the puppies," he said later.

Laying down the jackrabbit and seating herself on a wide base, Alatna bowed deeply aside, to one side then the other, in a perfect extravagance of self-humbling. She begged Cris in a voice so sweet and pitiful over the rabbit between them that, watching from the deck, I thought he ought not to go on but he did. "I thought she was just thanking me for the rabbit," he explained afterward.

Alatna sprang at his face and hit it with open jaws. Not daring to turn his back Cris backed toward the gate, pleading at every step. Three more times she hit him in the face. (But she did not close her jaws.) Then he was safely outside.

Now I had the fragment of the Rosetta stone of understanding I had wanted long before, when Mr. Arctic, standing beside Alatna and her toy, had whined to me with novel urgency. My puzzled yielding had spared me trouble. But that big young male had not humbled himself as Alatna had just now done; probably he felt more equal to the attack if it became necessary than the new mother had felt.

Crawling into the den as usual the next day, no sooner had I reached the middle chamber than Alatna, in the rear one, rose and came so swiftly around the obstacle rock, a fixed look in her eyes, that I had only time to crowd myself against the rock wall at my right. Despairingly I told myself, "You've gone too far at last."

Alatna scrambled up over my left side and hurried out along the tunnel. I was alone with her sleeping puppies! I cast one longing look at them. Now was my chance to touch their softness and to learn how many males and females there were. But if I was delegated as wolf-pup sitter, the job was one I wanted no part of. I backed out of the den as fast as I could crawl. Alatna was on the far side of the pen having a bowel movement.

The "second spring" of the Tarryalls arrived one day—the leafing out of the aspens. The next day the chartreuse leaves in the groves among the mountain bases were so billowing-light, so ethereal that they looked like clouds of perfume materialized. That evening in the den I noted one pup's eyes partly open, fixating the flashlight. The following day Alatna buried all food, not eating. This caching was like her devoted caching before giving birth; it meant change ahead, but I did not read the action.

Soon it was plain that Alatna was in trouble. I tried to fathom its nature. "She is trying to lead her own life," I thought. "Alien among us, humans and dogs, that shadowy animal down in the den has imperatives within her to help herself, defend and feed her young, whereas the dogs are dependent: full of dog-to-person needs, they live in a manipulated world." But the wolf felt responsible for herself.

Sometimes she looked away toward the empty hillside as if for help, as she had done during labor. "She has no wolf to aid her," I thought. "Her needs are unmet. In her aloneness she does as well as she can."

Only one fact was clear: Alatna wanted caches for her pups around her den. She could not keep the ones she achieved: the dogs dug them up. Besides, much of the food given her—eggs, kibbles, chopped horse meat—was thwartingly unsuitable for burying.

All one day, especially toward evening, she acted disturbed. Did she think to move her pups? But—to where? She was still restless the next day and now at last I saw what her real trouble was. The den was crowded. She was holding the puppies in it beyond the time when normally they would have been toddling in and out. Probably she feared the dogs on their behalf. She had an additional misery: she was keeping the den clean by the only means open to her, swallowing the feces of the puppies, as a mother dog may do.

The puppies were restless. That afternoon they came fearlessly around the obstacle rock into the middle chamber and drank from a pan I held for them. One puppy started to follow me out along the tunnel. Watching sharply, Alatna called it back, not with the puppy call but with a summons. It obeyed. After this warning call the puppies were more quick to shrink, to doubt my huge form; the least surprise move of a pan startled them.

All through this period Alatna was feeling her way in a situation not meeting her instinct. With watchful, improvised, hesitant conduct she coped with her captive condition. On the first day of June she again performed hesitant actions as if to move her puppies. She did much futile digging on a ledge near the den mouth.

Glancing out the door the next morning, I was bewildered to see unfamiliar brownish little figures toddling around the den mouth. The puppies! They were out. I hurried to warm milk with a beaten egg yolk and take it down to them. They clustered softly at the pan and lapped. I let my hand drift down onto the air-soft fur.

Meanwhile Alatna went to a cache she currently had and nosed it. Innocently I noted, "She wants to absorb the puppies' attention, feed them herself." No crossover occurred between my thought and my action: I stayed happily beside the puppies. Then Alatna blew, with closed mouth—a kind of snort. Brother! I moved off silkily.

That evening Alatna met my persistence once more. The puppies were in the den and I took them milk flavored with dog food. They met me in the middle chamber and drank. Presently Alatna, in the rear den, moved to come out. I backed out hastily on elbows and knees: I had been trapped once with her scrambling over me. But she simply stopped at the pan and drank up all the milk. I was disappointed but forgiving. "She's thirsty," I thought. I was wrong. A food war was on between the wolf and me, and this was to be a major tactic in it: she would eat up what I offered to her puppies.

I could have cried with disappointment when she produced her second main tactic. I had prepared a special treat for the puppies, finely chopped beef heart. They were eating it when Alatna took it from under their noses and buried it, pan and all, ruining the meat bits. If she could not choke down what I offered, she would bury it.

I was righteously determined to feed the puppies not only for my pleasure but because I was trying to regain their lost confidence. I had lost it for the reason, so I supposed, that friends, dear and welcome, had arrived at a critically wrong time—just when the puppies were emerging from the den and I should have spent all my free time accustoming them to me in the pen as in the den. But I know now that I should have lost them anyway: Alatna so decreed.

Among the four puppies there was only one female, Miss Katmai. Her brothers were Kobuk, Kotzebue and Kuskokwim. Dog and wolf characters mingled variously in them. All but Kotzebue would look like gray wolves, though Katmai had a thick muzzle, not a wolf-slender one. The gray-wolf puppies had cloudy gold eyes, neither the pure bright gold of the black wolf Lady's eyes nor the clear gray-green of Alatna's eyes. Kotzebue had dark dog eyes in a very fair blond coat. But in disposition he and he alone was like a pure wolf. It did not occur to me to wonder about the disposition of the others.

Kobuk, the darkest of the males, was choice of litter by Baranof's former master. I pupnaped him one morning, just as he toddled fast away from me into the den. Cris distracted Alatna by playing with her over on the bench. Cuddling the silent puppy under my jacket to keep him from crying and alerting his mother, I rushed him into the bedroom and then ran to call Cris. He came at once and drove dark silent little Kobuk away to his alien life down at Colorado Springs. Alatna did not seem to miss him.

A new trouble beset her. Bobby and Wallie, her devoted retainers, now became contenders for the pups' affections. They refused to leave the pups. Even when these nursed, the two intrusive dogs lay among them. Alatna's eyes looked stern and sad but the temperate wolf did not molest her dog retainers.

Her greatest trouble so far came on the day when she encountered the half-breed rebelliousness of her own puppies. It was the day when they discovered the cleft at the rear of the table rock. This cleft split the rock from top to bottom and from side to side but had only one entrance, among bushes at the back. It was so narrow that only the puppies could use it. They entered and curled down to sleep in a wide place near the closed end of the cleft.

Alatna tried to get them out. She stood at the entrance and cried the wolf puppy-call. It was disregarded. Using all her arts opportunist-wise, she tried to attract her puppies if she could not command them. She brought a worthless bone and cried. She brought an empty pan and cried. She dug, trying to widen the rock-walled entrance. She cried. She cried so much on this day that her voice became hoarse.

Once a puppy did come out. She put her jaws over his back and tried to walk him toward her den. He rolled over on his back and fought her—something no pure wolf puppy would have thought of doing. He returned to the cleft. Alatna, carrying her useless bait bone, went into the cold den and lay there alone. When I crawled in and spoke comfortingly to her she groaned.

But she was not licked—not yet. The noon sun shafted light onto the sleeping puppies' fur as they lay in the wide place. Alatna came around to the front side of the rock. Standing on a ledge she dug slantwise downward toward the nearby, unseen wide place. The sun beat on the face of the rock. She panted. Her tongue dripped. She cried. She kept on digging. And she opened a new entrance to the cleft, meeting the wide place on the nose as if by survey. Like the first entrance, it could not be widened. The puppies had a second entrance to their hideout, that was all.

Now Alatna was licked. Standing on the ledge she drew a deep breath and expelled it hard in a desperate, baffled sigh almost exactly like that of a human being who has worked hard in vain.

The next day she was resigned. Her hoarse voice was still. The puppies went into the cleft and slept, as they would do from now on until too broad to squeeze in. Alatna made no further effort to get them out. She had tried her best to keep the ungovernable puppies with her, away from the dogs, away from me. Once that day I stood with my head near hers as she herself stood on a rock, and I commiserated, "It's hard, Alatna." The wolf glanced at me and groaned a little and kissed me briefly.

Poor Alatna. She had tried so hard all along to make things come out as a wolf should have them. I felt sorry for her when as days passed she still seemed to have given up on her disobedient headstrong babies. But she did not stay beaten. Situations changed. She tried new things. And eventually a good change took place. The puppies, which had, distractedly, seemed to follow and "love" the dogs more than her, now were overjoyed when she came to them. They fell on their backs, patted her nose like little wolves, hurried alongside her nose as she walked. Was it because she was always loving, whereas by now the dogs bit them sometimes or stood stiff-tailed, growling at them?

The puppies still ran to Brownie, the only big male in the pen by day, flattening their ears sweetly. Brownie ignored them. Kotzebue would follow him, trying to throw himself at Brownie's feet. He hardly had time to get himself laid down before he had to scramble up and run to try it again, under Brownie's indifferent nose. Did wolflike Kotzebue more than the others miss a Father Wolf?

In July the cares and frustrations of captive motherhood began falling away from Alatna. As the puppies came toward her along the trail early one morning, Alatna, waiting for them, stooped her chest to the earth, hindquarters high, like a cat. She sprang at Kotzebue, her favorite, and hit him gently with her nose, then turned away with a spring, her forepaws spread wide. All the dogs and puppies gathered under the big spruces, full of play and gaiety in the cool dawn shadow. High above them a paleness of the first sunlight lay behind black trees, small with distance, up on the crags. Beyond that paleness of light rose other dark crags still in shadow.

Little earnest Katmai was hit by flying paws as Alatna galloped right over her and Katmai was tumbled and rolled. She picked herself up, pleased and distracted in the excitement, ears at distraction position, half back. She tried to foresee what would hit her next, yet not be out of the fun.

That evening as usual came an event of pure happiness for the wolf animals. It was the daily return of Baranof from the mine. He strode through the pens, making the tour of them, followed and adored by Alatna and her puppies.

One evening a last frustrating skirmish in the food war occurred. I had brought from the Springs an expensive treat, three beef heads. Cris chopped one in two lengthwise and took half to Alatna. She would not come to him. So he took it into the home pen, where the puppies were.

"Oh, let her give it to them," I called unwisely. "She loves to give them things." He brought it out. But when I essayed to lure Alatna into the pen with it, intending to shut her and the puppies away from the greedy dogs, she seized it and bore it away into the big pen, where the dogs got it.

At last each dog had a chunk and was gnawing. One of the younger dogs had Alatna's big chunk and his own. Poor striving Alatna had one small chunk by the travel box in the yard. Two puppies were with her and she tore bits off it, crunched them and gave all to her puppies. Little Katmai was alone in the home pen, gnawing ineffectually on a chunk. I was silent with disappointment. All our work and money only for this: the dogs waxed fat, as usual; the ones we longed to favor grew thin.

"Alatna feels her responsibility so heavily," I thought forgivingly. "She thinks she knows the right thing to do and she's going to do it." It was not apparent to me that another female creature was at hand, equally sure of what was right and equally determined to do it.

The food war waned; Alatna had won it. Often the puppies would step to a pan of meat bits I took to them, eat a morsel or two, then go away, though I had withdrawn. But if Alatna came up and bent her head to the pan even without eating, the puppies ate the meat all up at once, continuing though she walked away.

Rising late one morning I had much to do before I could feed the puppies and Alatna—fire to build, pan of eggs to bring from the storeroom, eggs and milk to prepare. Alatna came into the cabin, searching for something to take to the pups. She found nothing. Finally she took a peach from the table and carried that outdoors. The puppies trophied it first, then actually ate it. As so often, that okay sign from Alatna—her nose touched even momentarily to the doubtful food—meant Go ahead. They ate.

August was a time of liberation for Alatna. Ever since January, when she began her first den dig, she had been driven in one way or another by the cares of mating and motherhood. Now she was free. Old ways were resumed. She demanded imperiously to be let into the cabin. If refused, she went off and howled or else she tried to get in by her own efforts, smashing the glass with big drumming forepaws and biting the wooden framework. She hated dog fights as much as ever and ended one by a new method one day: she parted the dogs with difficulty and when they still stood and growled at each other she laid herself down between them. She was curious about novelties as formerly. She dismissed me shortly once and stepped aside to concentrate on listening to the rough sound of a small aircraft, approaching unseen back of the mountaintop. It came into view, a helicopter, Alatna's first.

Above all she hungered to travel again. I did not realize how strong this desire was.

"Lucky person that I am," I thought gratefully one morning. I was sitting on the hay-sweet wild grass beside the trampled play place of the puppies at the foot of the pen tower. There had been a wild play with Alatna when I first came over here to the bench. She had spread her paws, slanted her ears back, twitched electrically when I threatened moves. Then she had broken into running—around the big rock by the Wolf Den cabin, down the hill and around rocks there, her ears back, her tail flying. When she saw that I could never catch her, she took to dashing by so close that I grazed her hair with my finger tips. In this way she made the game more exciting for herself. By puffing and screeching I tried to add the excitement my slow feet could not give.

Her puppies had watched, their tails gaily elevated, too short-legged yet to join the run. But what chases should be ahead. Now in the still sunshine the puppies lay in dusty hollows under the tall fir by the play place, breathing in sleep, growing to be big animals. The squirrel chuckled. Alatna stepped up from behind and kissed my face for a moment, then was away, her restlessness not quite appeased by running. The restlessness did not trouble me.

A week later she left. With her went Bobby and Wallie, her devoted retainers, and the three big wolf-dog puppies. The puppies came home in the afternoon. I got them into the pen. Kotzebue had one porcupine quill in his nose.

Late that afternoon the heavenly, heavenly sight! The outside dogs barked; I hurried to look. At the outer gate was Bobby—Wallie, too, but dark against the dark cut bank she did not show at first. The next second, Alatna! Not outside the gate but inside the new garden pen. The agile wolf had leaped the fence. She cowered and hurried and cried along the fence to find a way into the yard pen, and came instantly to the gate when I opened it, though it gave only into the dogs' courtyard. I did not even have to stand aside. Once in the courtyard she gave a bad time to the two dogs that were there. They were of the litter of Gray Lady's that she had always hated and one of which she had killed. Especially she persecuted a young female named Prancie, biting on her, erecting mane and rump fur, pursuing and snarling at her.

As quick as I could stumble over dogs and wolf and throw open the gate into the yard pen, again Alatna rushed through. No matter of coaxing, getting meat bites, luring. She wanted in. That was where she had in mind going. Her surprised pups, who had moped all afternoon, flew to her, all soft fur, crowding around her, until suddenly the wolf bethought herself of something. She proceeded to bite on each pup and make it lie on its back, or cower, or run shrieking. Each muzzle she took crosswise in her own and chewed lightly on it. Why? Because—so I thought—these pups had deserted her on her trip. Because she felt irritation at having had to come home, as she would never—again so I thought—have done if all had been with her. At least not for a long time.

At this biting the pleased pups, who seemingly had to love someone, turned to Wallie and fawned happily. One pup turned back to Alatna and, not daring to kiss her face, kissed the air a foot in front of it. And the pups began to play.

When Kotzebue had returned with a quill, we knew the travel party had met a porcupine and that some or all would have quills. Also that probably the reason why the pups had left the party and come home had been precisely the porcupine—the commotion and tumult it must have caused. Alatna, wolflike, had not a single quill. Bobby's shoulder, jaws and inside of mouth bristled with quills. Wallie had a few; hers would have to come out of themselves; we could not touch her. But poor whimpering Bobby I cuddled and commanded while Cris jerked out all her quills.

September was a month harrowing with escapes. The old fences had never been repaired; I still toured them constantly, watching for escape signs. The final escape occurred on a Sunday luckily, so Cris was at home. Alatna must have gone out swiftly, digging under the fence to join Baranof, who was outside the pens. I had made a fence tour only half an hour before.

It was Bobby that let me know about the escape. She appeared at the door alone, panting and crying. I ran to find the escape place, then ran to tell Cris, who set out at once in the pickup to hunt the animals. Alatna loose was no menace to man or beast, but Baranof was a different matter: free with companions he menaced cattle.

Wallie soon showed up, so now just Baranof, Alatna and the puppies were loose.

Cris came upon Baranof in the act of biting a cow. He

leashed the dog and with him as lure was leading the whole group homeward, Alatna happily following her mate, when Victor dashed up in a jeep, putting all the wolf animals hopelessly to flight. (People seldom realized the limitations under which we operated, to win the cooperation of a wild animal.) Cris brought Baranof home. Now there was nothing for us to do but wait for Alatna and the puppies to return of themselves.

In the afternoon the three puppies showed up but only pretty little Katmai chose to come in. Her brothers went on away. At the gate Katmai courted Bobby, lying down before her and softly raising huge paws. Then, endearingly, she hurried to the home pen and "loved" and bit and played with the three baby dogs that were ensconced there. By "bit" I mean that she put her jaws over a puppy's back as a wolf does when fondling or walking a puppy. I thought it was to play with the baby dogs that she had chosen to come in. She was unhappy though and put her head back and howled.

The arrival of visitors precluded hope of Alatna's return; she would never come back while strangers were here. We went to bed lonely that evening, as were the pen dogs and Katmai, listening, not sleeping. In the morning Cris announced good news: he would stay at home from the mine and repair the fences no matter how many days it took.

I set out for the Springs to buy rolls of heavy fencing to widen the ground apron around the foot of the fences. Even if we only had Katmai to hold, we wanted to make the pens hold her. On my way I stopped to tell Enid the news about the fences. She had heard my car coming and stood waiting at her gate with a face of doom. (It was one of her cattle that Baranof had bitten.)

"*Now* what?" she inquired grimly. The last thing she could have thought of was that I bore good news.

I told it, then went on my way still troubled: Alatna and the wolf-dogs were still out and liable to be shot. All that day, running errands in the hot city, I thought anxiously about them. It was dark when I arrived home. Alatna and her young were safe: they had come into the pens that morning soon after I left.

The next day, in the lightness of my relief over the fences and animals, I looked with new eyes at the big young wolf-dogs and saw them as if for the first time. Kuskokwim and Katmai came into the cabin as I mixed bread. Gaily I talked and talked to them and once Katmai's back rippled as if she were being petted, yet I could not touch her nor her brothers.

I came home from the dog walk that afternoon—among bright-dark aspens, gold and green, and blue mountainsides—to find Kuskokwim in the home pen with the three baby dogs that currently lived there. He was holding a rope in his mouth and gently walking, followed by the puppies, their little yarn tails wagging.

Like pure wolves, he and Katmai were born puppy-sitters. They were more sweet and devoted to baby dogs than some girls are to human babies. They sought them out, lay and played with them, watched them. Trigger and Lady had been more than a year old when they adopted the five wolf puppies, so we had not had anything to reveal to us that young "wolves" would be so enamored of puppies until now, when Katmai was their happy slave and Kuskokwim played games with them. And I thought grave wolflike Kotzebue was indifferent to them!

Returning from a drive on the day the fence repairs were finished, I brought a highway-killed jackrabbit to Alatna. A baby dog pup was barking incessantly in the home pen; I had forgotten to feed the puppies before leaving. The puppy fell silent and at the same time there came a strong wolf puppy-call—from Alatna, I thought. But it was from Kotzebue. He had got the rabbit and like a wolf had brought his stomach-load of meat to the babies. Afterward he lay and played with them.

All through warm sunny October my halcyon feeling of liberation from fence cares persisted. Once I even took a long walk all by myself, leaving every wistful dog at home. I sauntered and gazed instead of hurrying to keep up with the dogs and control them.

But on a November day of low dark clouds the awareness caught up with me: those escapes had meant that Alatna was not happy. Having puppies of her own had not put her in paradise after all. I wished for a sign—a sign of whether Alatna, who had dug and looked away through the fences at freedom so near, could ever find, say, 51 percent of happiness in her life to make it worth living. I watched closely for a sign. It came.

Near sundown the wolf animals had given me a good scare by apparently disappearing from the pens. With bitter resignation I was trudging the rounds to find where they had escaped when, looking upward, in the late low sunshine from under clouds I saw fair Kotzebue standing on the hillside below the bench. I hurried up to him and here were all the rest—Alatna, too, with her ears flattened down. I hallooed and sang and teased.

And the sign was this. Alatna lay down in the last sunlight crossing above the woods in the draw. She yawned. And at the peak of the yawn she shook her head a little, as a happy dog may do, in high good humor. I ran back to the cabin, blithely unaware that no doubt my own ebullience at finding the animals safe had elicited the sign of good cheer.

13 *Changes*

ALATNA was exhilarated by the first snow, as she always was. Change stimulated her. It was Bobby's barking at her as she flew over the still-dark rocks in the home pen that attracted my attention. I was washing dishes and did not see how the wolf-dogs took their first snow. But the next day they met their first ice like pure wolves. They were so interested in examining it that they could not take time off to come to breakfast. They were at an overflow pond in the brook, looking down at the ice, rearing back and drumming on it.

On a sunny December day of melting snow I sat writing Christmas cards and they were strewn all over the table. So when Alatna, after rolling on the cake of Ivory soap, leaped onto the table to caress herself on the window as usual, I got her off—also as usual—by rushing outdoors in great commotion, causing Swagger to bark excitement. I snatched and dipped in water a sliver of soap as I went, and as, sure enough, Alatna emerged, pushing open the heavy screen door, I flung the soap down just as she bent her head. So it struck her muzzle by accident. A fleck of suds must have gone into her eye for she acted for an instant as if she did not know what had happened to her.

She struck across her near eye with her paw. I knelt lamenting. She kissed me and bit Kotzebue. All was confusion as she half closed that eye, kept returning to me to have it "loved" and turning away growling to take out the puzzling hurt on the nearest dog or wolf-dog.

One cold moonlight night—it was while Bobby and Wallie both were in heat and Baranof was staying in the pen nights but rejecting Alatna—I could not sleep and finally I rose and started toward the cabin. Up onto the plaza in the moonlight to meet me came Alatna, followed, inevitably, by a small dog or two. She whimpered to be made a fuss over. I loved her. She went to the door, in black shadow. But I did not want her indoors as it was cold and I had to make a bed for myself on the bunk. So when she turned toward the gate I whipped in at the door and trying to close it felt a heavy object—Alatna, caught, in the darkness, between door and screen.

So in she came and got onto the bare mattress. I knew right away what her plans were when she turned and turned. Sure enough, she lay down in a curl. I knelt on the bunk and put my arms around her. She was luscious to cuddle, more yielding in the dark. If she could not have Baranof, at least she could sleep here in the cabin. But I was freezing. When I unrolled blankets beside her she left.

Baranof was unfaithful but from her point of view he was still her mate. At feeding time the next evening Cris tried to induce him to leave his two bitches and come indoors. Finally Cris scolded and at this, sailing up onto the terrace and then the plaza, came Alatna, black-eyed, ready to take charge of the situation. She looked with anger at Cris and he edged toward the door. Then—and this is the point of the incident—Alatna turned toward Baranof and kissed and kissed him, as if to console him for that damaging tone of voice Cris had used, which had affected the thick-skinned dog not in the slightest degree.

One day at the close of December there came such a prolonged squalling from a dog pup, undiminished, that I went to see what was the matter. (I always saved myself trouble by waiting to see whether cries would go diminuendo of themselves.) Alatna was in a state of emotion. Baranof had bitten the pup and she wanted to punish him for the deed. At the same time she did not want to fight her mate.

Fighting she was at the moment though. Both animals were reared up snatching at each other, Alatna's eyes black, Baranof's muzzle bunched to bare his fangs. After a moment Alatna ran to the cowering puppy, touched it with her nose and then ran back to lunge at Baranof. Her registering of ambivalence—intensity of drives two ways—was an exciting thing to witness. She rushed to one, then the other, intense feeling showing in her black glittering eyes—rushed to rear up and gnash crying at Baranof, then rushed to nose the pup, sitting to one side, its prolonged screeches now subsiding.

Two weeks later she behaved in a different fashion toward her dog mate. Significantly in playing with him she always lowered her height, either by bowing her chest or by tilting her head below and aside from his. Once she lay clear down on her belly. As she started to rise she made a stirringly lovely gesture. In the flash of a second she reared her head to one side and curved and reared it to the other, all in one sinuous movement, so gay, so graceful and vital—and always with that accompanying eye expression, black and electric and gay—that a dancer would have longed to rival it. Alatna was in estrus, her second with a mate.

On this same day I first noticed a strange new theme sounding its opening notes among the varied themes of vitality at Crag cabin. "Kuskokwim looks dreamy," I noted, "perhaps because his eyes never light up with love toward a human being. Perhaps for other reasons. Possibly because of some 'heaviness' connected with being a dog and a wolf and yet living the life of neither. Neither pet-friend nor wild."

Before Alatna's puppies were born I had felt uneasy about living with wolf-dogs because I had heard they were dangerous as wolves are not. But my first sight of the soft wild-wolf-colored babies in the den with Alatna had banished my fear.

The thought of possible danger had not entered my mind again.

"Sweet he is," I concluded my meditation about Kuskokwim confidently. That morning I had given each animal a big chunk of cornbread. The others ate theirs. But Kuskokwim had held his chunk lightly in his jaws and, glancing around, had started off with it. He had not gone ten feet before he practically "smiled." A baby dog pup had found him. It reached up to touch his mouth. He let go of the chunk, gave the pup a brief kiss and stood by while the pup started to choke down the cornbread. Patiently I gave him another chunk. He did the same thing with it, except that it was for the other baby dog that he let go of it. I had one last small chunk. I gave it to him and he ate. Both puppies were still gnawing on his gifts.

He had drooled a little with one gift in his mouth. "That's self-control," I thought approvingly, "to choose to give and not to eat."

One cold February morning all themes were sounding except the mysterious new one of wolf-dog nature. In Rites-of-Spring turmoil I was turning out a bit of writing called for in the mail brought home the evening before. Alatna, still in estrus—the long estrus of a wolf—but not yet at its height, was pounding at a door pane to get into the cabin; if necessary she would smash the pane as usual. Swagger was crying to get outdoors to her, though she would have killed him with pleasure if he had started any hanky-panky with her. Baranof was barking to her, Stay away from me, woman.

The dog Prancie, the one that Alatna especially hated, lay on the bunk nursing her first and only born, a solitary male pup, and Wallie, whose own newborn puppies had been killed by Alatna and who was still hemorrhaging, was shrieking to get to this lone pup and mother him. A storm was forecast and Cris was preparing his own breakfast at the small stove by which I huddled writing, in the still-freezing temperature of the room. He was hurrying to get off to the mine and load the quartz blasted down the preceding day, before today's storm should cover it. He would mail my script on his way. Each theme came in clearly and I smiled with pleasure. I delighted in the "itselfness," the yon-selfness of each living creature around.

A week later Alatna reached the height of estrus and stood. Baranof stayed at home from the mine. The top half of the big, once-handsome door was boarded up, its entire inner framework and all glass smashed and gone. Prancie was dead.

We had driven to the Lake one afternoon, leaving her and her pup asleep on the bunk. On our return I went to the cabin to build a fire and start dinner. At the door I stopped dumbfounded. The top half of the door gaped. Its inner framework lay broken on the cabin floor among shattered glass. Cold wind poured through the cabin from the escape crack.

Prancie came with strange joylessness from the bunk nook to meet me. I saw what had happened to her. A lump that had been on one side of her neck had shifted over her throat and was strangling her. Otherwise she and the puppy were unharmed. Alatna must have broken the door to get at this dog she hated. Prancie was a timid dog but she must have braved Alatna off. In her agitation that lump had apparently moved. She suffered for a day, then was given the mercy of instant death. Vets were still strangers to our life.

We supposed that the pup, Timmy, must die too. But we decided to give him one chance. He was carried to the bedroom that night, where he comported himself so debonairly that with amused pleasure the decision was made: Timmy would live.

Soon he began refusing the milk I had been spooning into him. Bobby, sterile from birth, was giving him milk. Confidently the pup had applied himself to the correct locations. At first Bobby froze in timid ecstasy. Later she became the most blissful of dogs. Twice she was obliged to withdraw because of inflamed breasts. Aside from those times she nursed the puppy through to weaning time, then weaned him authoritatively.

An iron framework was ordered for the door, to foil Alatna. A bit sheepishly we made one proviso, that the pane above the doorknob, the one Alatna had always kept smashed out, was to be omitted. We had felt abused because of that "wolf hole," but we found we missed the big warm personal head intruded at the opening to observe and commune with us.

During the pause of Alatna's pregnancy when outwardly all seemed to run along as usual, I flew east briefly for a speaking engagement. Cris met me at the airport on my return. It was dark when we drove up our own road in the draw.

"Call to Alatna," urged Cris. "She's missed you so."

Gladly I rolled down my window and shouted into the night. I saw the tossing eyes at the fences and knew the animals were barking or howling wildly at hearing my voice.

But then came an awkward pause while I washed up in the greenhouse and changed clothes. My seatmate on the plane had suffered from a truly vicious cold. Perhaps my precaution was useless but it was the least I could do for these animals of ours, none of which had received inoculations against disease except Baranof and Alatna, and she only for rabies, not for distemper. Their only safeguard was their remoteness.

The animals fell silent; they must have gone back to sleep. The courtyard was dark and silent when I opened the back door of the bedroom and crossed on noiseless soles to the side gate by the cabin. As I felt for the bolt, from the blackness right beside me, on the other side of the gate, came a sound like none I had ever heard before—a kind of strangled howl or broken-off groan, deep and abrupt. The next minute I was inside the pen and Alatna, an armful of fur, was standing up in my arms, kissing and kissing me. She must have been crossing the deck from one side gate to the other, trying to determine whether I was really at home or not. That choked groan was one of the most heart-warming welcomes I have ever received.

She could not get enough of pinning my shoulders with her forepaws and kissing my face. Off she would go—the yard light was on by now—to mingle with the welcoming throng, but she had to hurry to me again and again to kiss once more. She nearly knocked my teeth out, jumping to kiss me when I was not looking.

Alatna denned one evening in late April. Bobby and Wallie, her loyal followers, who usually slept against her fur down in the spruce shelter, slept for the first time up on the deck. A dog shelter had been made there by walling the underpart of the deck table and bedding it with hay.

In the morning the yearling wolf-dogs stayed down under the shelter even though Baranof entered the yard. After Cris and Baranof had left I noticed all three wolf-dogs coming out of Alatna's den; they had paid their mother a visit. But far from wanting or permitting Wallie to couch against her as in her first labor, Alatna growled when any of the female pen dogs offered to go into the den. Each retired immediately and did not try to return.

The wolf-dogs went up onto a high rock in the big pen and had a howl all by themselves. This was an unusual proceeding. They were lonesome and at a loss without Alatna. Not until now had it been apparent how much they depended on her for leadership and confidence.

I cheered them for a little while with tidbits. In the afternoon I started to hunt them up, for they were not lying in the shelter though the day was raw and cold. I found them all lying as close to their mamma as they could get, namely, on the very nearest bit of unwet ground, the uncomfortable slant of the foot of the terrace, a place where they never lay. Those lonesome yearlings! They were listless and depressed all day. Not until evening did they pluck up heart to play.

The next morning they were overjoyed to see Baranof. Never had he received a bigger welcome. Tall bodies stood up on the inside of the fence as he strode toward the gate. Jaws opened wide to wow with grateful relieved welcome. The excess of this welcome over their usual one was the measure of their inner lostness.

Baranof hustled down to the den mouth in his usual big-shot

bluff way, shouldering past the crowding wolf-dogs and dogs. Alatna came out of the den to meet him. Her tail was clamped down, her hindquarters were shuddering. But her eyes were kind and radiant and she was crying for fellowship, not from pain. (Her labor had begun at dawn.) That was something to see—the animals thronging the emotional wolf who was in labor. She could hardly get over or past the throng into the tunnel. She did not go into the back chamber but turned in the middle one. Wallie and Bobby now crowded in with her.

Among the yearlings Kotzebue showed up as her favorite, as usual. Katmai was staying away. Kuskokwim was afraid to venture into the den; he looked in at its mouth. But Kotzebue went in, sparkling-eyed. Alatna took one of his front ankles in her jaws and held on quite a while. It must not have been a hard grip; he did not cry. All surged out, including Alatna, and she reached for his ankle again. He looked happy, wrestled her head. She seemed to like the close body, the contact, the attention.

Abruptly she changed. Her fur foamed up. Fierce, brilliant-eyed, she rode Kotzebue and reached over his shoulder to bite at his throat. He walked slowly, bearing part of her weight. He looked miserable. Startled, I realized that probably this was what I had escaped, the year before, by my sudden immovable determination not to go near Alatna at the close of her labor.

After Cris had driven off to work and I was alone in the cabin, I thought I heard another kind of cry, not the pain cry but a baby cry, and I stood on tiptoe to look out the opening at the top of the still boarded-up door. At the same instant the first sun over the high crags dazzled me and a wolf baby cried.

By a quarter to nine the pain cries, which had become frighteningly hoarse and harsh, ceased. I took a bowl of fresh milk into the den. Alatna lay with her back toward me, busy taking care of herself, cleaning her rump and bending her nose to the babies I could not see. She drank the milk willingly, though as a negligible item. Her mind was on her totally absorbing affairs. She had to turn her nose over her back to drink as I rested the pan on her fur. So milk was slopped. I withdrew the pan and reached my hand past the obstacle rock to mop her fur, but she said no as definitely as if she had spoken human language. She pressed her nose to my hand for an instant. Not hard pressure, just No.

Two days later she came out of the den, re-established authority, tented her ears kindly at each noticing of me, accepted and seemed to expect the smelling salute from the female dogs.

A new era was beginning for Alatna. She was regally the matriarch. Young life, soft, boundless and unconserved, surrounded her. The year before, when I had expected her to be blissful, she had been filled with anxiety over her puppies. The year now ahead was to be one of troubles, some of which would alight on Alatna. But through them all she would be happy.

One of the first troubles was that I caught Colorado tick fever. It is not lethal like the Rocky Mountain tick fever but it is incredibly debilitating. All summer I dragged about my work. For a long time I did not go over to the bench. When I did go to the far end of it, I stopped in amazement. The dark woods were aglow. All through them floated big knee-high yellow daisies. I looked for that pretty show the following year but it never came again, though other wild flowers bloomed. This year everything suited the daisies; conditions were just right.

And all through the year of troubles that now lay ahead there would be the gleam of Alatna's romping spirits, like those golden daisies in the woods. She would have more cause to be troubled than we did, but when she ought to have been downcast she played. Her spirits were high. She was her old gay tundra self as she had not been since the loss of her fellow wolves.

I did not wonder what had changed in her life to cause this gaiety. I think now that it was due to a combination of things. She had baby puppies, a happy fulfillment for her, and this year she did not worry about them but knew what to expect. Moreover, she had the daily companionship of the adult wolf-dogs, creatures more or less of her own kind.

14 *Distemper*

One cloudy changeable morning in late May old friends had just driven off after a two-day visit. The last wave and honk from the disappearing car and I hurried to the plaza, calling the wolf crew. They were all waiting and emotionally ready for the outlet of a howl. Alatna, who had not come up to the plaza the whole time our friends were with us, bounded up over the terrace and greeted me with tented ears and shining eyes. She kissed my face, jawing it and wowing at the same time.

Then the three wolf-dogs, who had sat at the head of the trail across the brook, wrapped in attention to all that was going on—the departure, the demeanor of Alatna—were convinced that a howl was starting and came running down the trail to join in. Alatna stood on the terrace, pouring head and voice to the sky. Below stood the others.

Next came lavish pans of milk—fresh, not canned—for all, including the baby wolf-dogs down at the dark rock den mouth. I held little Wildie, the only female in this year's litter of five, in my arms for a minute, hoping she would be "my dog" someday.

Already I was losing this year's litter as I had lost the first and for the same reason, so I supposed—friends arriving for a visit just as the puppies began emerging from the den; I had been too busy to cultivate their friendship. But for a while yet I would be able to cuddle the huge soft shy babies. We found homes for two of them. Left to Alatna and us were Wildie and her brothers, Doonerak and Mr. Shy.

Alatna played more with the males than with lone little Wildie, and she preferred Doonerak to Shy. Often she pinned Shy down domineeringly.

In early July we had distressing news. The master of one of Alatna's puppies had been jailed following a drunken shooting brawl, and the neighbor to whom the puppy had been taken for care, angered by some puppy mischief, had hauled him to the western slope of the Rockies and thrown him out to die. The master, so it was reported, wept in his cell on learning his puppy's fate.

From Lake George we telephoned a reward ad to the newspaper of the town nearest where the puppy was said to have been ditched. Soon a reply was received: such a puppy had been found. Early the next morning I set out in the bumpy pickup for the long hot drive to the western slope. I dreaded the day: I was still weak from tick fever; but I was glad for my errand and hoped it would prove successful.

At burning noonday I found and identified the puppy, at a house near the foot of the town dump. I paid the reward and with the puppy beside me in the stifling cab of the pickup I started the homeward drive. Hungrily the puppy ate a chunk of beef heart, still cold, that I had brought from the deep freeze at home. Then he laid himself confidently across my shoulders and looked out the open window. His confidence was a delight after the shyness of Alatna's other puppies. Finally he composed himself on the seat beside me to sleep, first laying one huge forepaw on my thigh. It slipped off in his sleep. He awoke and replaced it, then slept again. Not once during the drive did he lose actual physical contact with me. He had an ugly, healing gash over one eye.

It was warm dusty sundown when we drove up onto the roadhead at home. The puppy was daunted by the leaping, howling, barking lineups at the inner and outer fences. Luckily a little black "outside" puppy adopted him as a heaven-sent playmate and distracted him.

We did not return him to the wolf pens. We had never had a gentled pup of Alatna's before and we kept him for ourselves and named him Traveler. He and Alatna met in the cabin; they caressed but did not seem to have their hearts set on staying together. Traveler slept on our bed, growling imperiously if wakened and made to move over when we came to bed. He went along on the dog walks. His sense of smell seemed keener than that of the dogs. Perhaps the difference was that, wolflike, he disregarded nothing, noticed everything. At home he romped with the black dog pup.

Pleased, we told ourselves, "He's having the fun a puppy is entitled to have."

On the walk one August day Traveler ran off by himself through the shoulder-deep meadow grass, following a butterfly that wavered ahead of him in the sunshine. Afterward I was glad he had that happy special run.

That evening he left his pan of supper and ran in circles, staring blindly. We thought a bone splinter had lodged beside a tooth but on looking found nothing. He seemed unwell the next day, so at walk time I left him in the cabin. Standing up on the couch he looked out the high window at the happy dogs about to depart and he cried.

I rose at five the next morning to take him, too late, to a vet. He was dead. Cris had shot him. "He was agonizing," said Cris briefly.

We still did not know what lay ahead. All the animals were getting a discharge in the inner corners of their eyes, we supposed because of pollen irritation. But finally, taking Swagger along as a sample, I drove to the Springs to find and consult a vet. The day of veterinarians had come for us at last. I drove in my own car, a good used Plymouth. A car of my own was a major liberation in my life. I was about to need it.

The first vet I visited looked coolly at Swagger and refused the case, for the good reason, so I supposed, that the man's ankle was in a cast. (But no vet, ever, would come to remote Crag cabin.) That refusal was my good fortune. The next vet I stumbled onto was to become my wise and kindly counselor in the trouble now upon us and in future troubles.

Swagger had distemper. So all our animals had distemper. There was no time to lose. I was to bring as many dogs as possible the next morning for injections of live virus. The vet would teach me then how to give Alatna an injection. For the wolf-dogs, which I could hand-feed but not touch, there were white tablets of antibiotics. Dosage was by weights and I could only guess at those.

Driving homeward, flushed with anxiety and the heat of the day, I left the highway and drove the steep half mile up to the mine to tell Cris the bad news. In return, he told me that a *Life* photographer, Fritz Goro, was waiting nearby for a guide to Crag cabin. He wanted pictures of Alatna, a tundra wolf, for a cover story on Alaskan wildlife. Some visitor to Crag cabin had told the Time-Life office in Denver about her.

In his own car, packed in orderly fashion with photographic gear, Goro followed the Plymouth to Crag cabin. The day had darkened but with impressive care he prepared for the next day's filming. He went down the avenue with me, taking meter readings in different settings. Alatna fled, but his one glimpse of her showed him that he could not use pictures of her, not because of her eyes—I had bathed them as directed by the vet—but because of her crippled ankle; a second pregnancy had done it no good.

Goro took one picture it would not have occurred to us to take. It was of a pygmy forest of stakes planted around the mouth of a potential escape dig. To us it was only one of countless daily precautions we took to guard the wolf animals from escape and death by shooting.

Goro stayed at the cabin that night. (With the stranger there Alatna would not stay in the luxurious hay shelter under the spruces; though she was ill and the weather rainy and cold, she and her coterie slept for that night under Coonie's old tree, at the far end of the pens.) In the morning, using the Wolf Den as a blind, Goro tried for wolf-dog pictures through the open top of the Dutch door. Alatna stayed hidden but one wolf-dog kept coming around belligerently, as if on guard duty. After two cold, patient hours, Goro blew the whistle lent him as a signal and I went over and escorted him back to the cabin. Now for the first time I told him the story of Traveler and the distemper.

Angrily Goro flashed out, "Why do you give Alatna's pups to people that get drunk and shoot each other up? Do you think you are the only people in the world that like animals? Lots of folks would like to have one of Alatna's pups."

Humbly I supposed this was so, our difficulty being that only the relatively few local people were accessible to us. It did not occur to me to consider the still unfinished theme of wolf-dog nature.

Before leaving, Goro more than made amends by offering to help me with the dogs at the veterinary hospital the next day. He was to stay at the Springs that night. In the morning he was already in the waiting room when I arrived at the hospital with the pickup load of dogs, and when my turn came he accompanied me into the operating room, staying by the whole time while the dogs received their injections and I my coaching. I felt sure it was due to his kindly presence that I received the tireless patience of the two vets. Also they were interested in a case involving a wolf. Finally we left, I to save Alatna's life if I could. I believed I could.

At home I built a fire and sterilized a needle and the syringe and loaded the latter with costly live virus. Alatna lay relaxed on the bunk. Tremulous but confident I sat down beside her. I jabbed her. In one fraction of a second she dipped her nose to my wrist, brushed it as lightly as a butterfly's wing and sprang to a far corner of the bunk, where she crouched watching me, the needle broken off inside her. The virus was wasted.

I sterilized another needle and reloaded the syringe. Alatna was again lying at ease on the bunk. This time I tried to distract her attention. I set a dish of ice cream by her nose; I talked and yodeled and wooled her neck. I stabbed. She broke the needle outside herself.

I was shaken. This was not as the vet had promised things would be. He had said she would not mind the needle. But clearly it was one thing to give an injection to a dog mounted insecurely on a high slippery table in strange surroundings and another to give one to a wolf at home, with nothing on her mind but how horrible it was to be stabbed.

I tried a third time. She bent this needle double. I had to recognize that I could not give her an injection and that my failure might cost her life. I was the only person she would permit to handle her. I was in despair.

Alatna gave herself up to gaiety. She brought a booklet to the bunk and began tearing it up. I summoned myself to treat the affair as a big joke. Instantly she rolled over playfully and waved her paws. I petted and curved her body and she leaned up her face to kiss mine. (The broken needle inside her never seemed to cause any trouble.)

In the morning I drove down to consult the vet. He prescribed the white tablets for her according to her guessed-at weight. I placed no faith in the tablets. "Are these a placebo for me?" I inquired distrustfully.

"They will help some," he assured me gravely. "Besides, it's all you can do for her now." He added that it might be four months before we would know which animals would live and which would die.

Now I was nurse to a penful of sick wolf animals. My first task was to find a medium in which to give the white tablets. It had to be something liked by all that would be swallowed without chewing. I tried everything from Braunschweiger sausage to blue cheese and whipped cream. Finally I hit on vanilla ice cream.

The weather had turned hot again. Three times a day, armed with a chart, the white tablets and a carton of ice cream encased in wet newspapers under dry ones, I trudged uphill and down, hunting the animals. They would be lying wherever seemed to them coolest. Gently I followed each in turn, coaxing and holding out in my hot palm the loaded handful of ice cream. Only Alatna took all the tablets readily. On balance I got about two thirds the desired number down the other animals. I felt discouraged but the vet, looking at my charts, said I was not doing too badly.

All noses were rough and dry. It would be a long time before the black velvet ice was restored on them. Alatna's nose was the harshest one of all. Yet she remained radiant and authoritative. I went to the bench at late twilight once, purely from affectionate concern, not to give medication, and found her and the crew lying in a new place, on the side hill. Up she got, curveted once to be petted, then, her eyes gay with mischief, she tore off through the woods while the staid dogs looked on. She was more gay than the dogs, not aggressively demanding like them.

"That's something obscure and basic," I reflected. "It's the wolf personality. Perhaps the secret of it is that the wolf meets you without demands. Independent as a multimillionaire he meets you with soft ease, like a royal friend, demanding nothing, sharing and enjoying fun and friendliness."

Alatna was so much the leader that even the moods of the others responded to hers. If she was gay, their eyes turned gay; they trooped and tossed with her in gaiety. If she "talked"—and she was the sole one that truly, wolfishly did—they uttered their own indescribable ows and wows. Would they ever troop and "talk," happy-eyed, I wondered, without her to set the pace.

Bobby had been banished by Katmai from the pen and now met Alatna, her "owner," only in the cabin. Rejoined to her on the bunk, she would lay court to Alatna's face. Alatna might growl, Don't disturb me, and Bobby, crestfallen, crept away to lie in a corner of the bunk. Or Alatna might kiss her seriously but briefly.

It did not worry us that once-submissive Katmai had suddenly turned fierce toward Bobby: we had always said Bobby would get her due, someday, she was so mean and domineering toward the wolf-dogs. We did not think of the change in Katmai as one of the troubles of the year.

I washed dishes and scrubbed the floor all softly one afternoon: Alatna lay sleeping on the bunk. When she got up finally and started toward the door, Kuskokwim began to "talk" with joy. He stood on the deck table, where he had lain the whole time she was in the cabin, and he looked in through the screen door. He did not glance at me; his eyes were fixed on Alatna as she crippled out. Then he and Doonerak crowded against her shoulders. He had howled lonesomely a time or two, as if enjoying the purity of his own voice, when she first came in and lay down.

"I've never seen an animal so much *loved*," I thought. "Bobby fawning on her, Kuskokwim and Doonerak waiting for her, all the rest ready to rejoice or sorrow if she does, and be glad to be near her. Authority!"

Another face of her authority showed one night when a pup of Wallie's came down fatally with distemper and wakened us by crying. It was shut in the home pen with its fellows for the night. Nothing could be done for puppies so young, their eyes just starting to open. Cris went down to give the only mercy possible, quick, unfearing death. He returned to say that he could not enter the pen on account of Alatna, who had dug a tremendous hole by the gate and gone in to the crying infant. So I went down and felt around by flashlight and found the puppy. Alatna was curled near, helpless to do anything for it. But she was staying.

"Wolves are wonderful animals," I said, as we had said a thousand times, ever since knowing our first wolves, Trigger and Lady.

One wolf-dog, Wildie, had received no medication whatsoever. Katmai had turned against her as she had against Bobby and was keeping her away from all food, including the ice cream containing antibiotics. I was troubled for Wildie, but not troubled as yet by Katmai's change toward ferocity. Wildie was weakening; her hindquarters dragged. I could not help her. Even when Katmai was not near, she moved timidly away when I offered food.

A disturbing event was due on a Sunday in mid-September, the visit of an outdoor club. We had given permission for the visit long before distemper struck. At twilight on Saturday evening I went to the bench to check up on the animals. All were asleep at the near end of the bench except Wildie. Silently I walked along looking for her.

When I saw her I stopped in surprise. For a minute I thought she was playing. She lay on her back in a shallow pit beside a big flat-topped rock and bit and pawed like a playful grizzly cub at a dead aspen leaning across her. Then she whimpered to herself and I realized that she was helpless. She had fallen from the rock and could not get up. In spite of the darkening light, magpies were watching her, sitting above, hopping on trees and talking. She could only sit up like a poodle, helpless and terrified, when I approached and lifted her. In spite of her starvation she was a heavy burden. I carried her to Coonie's old tree and laid her there on the duff. I brought food but she was too terrified to eat. I left it beside her.

At dawn, knowing that now would be my only chance during the day to tend her, I went to the tree. She lay as I had left her; the food was untouched. She rose and staggered away from me, but I overtook her gently and gathered her in my arms as she struggled to climb a steep bit of path between high rocks. I knew that if I wakened the sleeping animals they would jump at me and make me drop her. So in spite of her weight I carried her the long way around—out the drive-in gate at the lower end of the avenue, up the road and into the bedroom, where I laid her on our bed. I brought food and coaxed her into eating a little. Then, leaving food beside her, I hurried away to prepare breakfast and ready the cabin and pens for the visitors.

We expected perhaps a dozen people and about four cars. More than sixty people came. Car after car, engines in low gear, roared up the road. Cars jammed the roadhead and overflowed up the jeep trail toward the ridgetop. Already the wolf animals could be glimpsed now and then, speeding in terror past trees on the bench.

Before admitting the kindly mob to the pens, we asked them to walk quietly past the curtained plastic-glass windows behind which lay the sick wolf-dog. This they did. We also asked them not to go beyond the spruce shelter in the big pen: the wolf and wolf-dogs required some measure of privacy. The people swarmed onto the plaza and down into the home pen. Cameras in their hands, they climbed the table and walrus rocks. They began slipping past the shelter. We checked that. Otherwise we were concessive. We answered questions cheerfully, in spite of our unhappy awareness of what this visitation was costing the animals. We hoped as always to make friends for the wilds.

But when at last the woman leader of the group remarked, "Your animals are playful," I blurted out the truth helplessly: "They're hysterical with fear." She gave me a serious, surprised look and soon afterward began marshaling the people for departure. This time they chattered noisily as they passed the bedroom windows. Out on the roadhead engines roared again. Cars backed cautiously down the steep, winding road. One car had to be dragged sidewise from its perch up the jeep trail. All was bedlam.

Silence at last. I went into the bedroom. Wildie was not trembling but jerking convulsively. I did not think she would live. But I induced her to drink a little warm milk. That night we laid her on a sleeping bag beside the bed. The next morning after breakfast I found her tearing up the bag in her terror. I carried her into the cabin and laid her on the couch, and there she stayed from this time on.

Wolflike, she was naturally clean and "housebroken." Several times a day I lifted her onto the table and before she could roll off I ran outdoors, wound open the escape window and lifted her down into the courtyard. Household movements and noises terrified her at first. She would rear her slender, pretty head, her eyes dark with apprehension, and strike a forepaw helplessly on the couch.

Gradually she became used to the situation. She queened it from the couch; that was her castle. Alatna often came in and "talked" to her and was kissed by her. Doonerak came in, climbed onto the couch and ate her food and played with her. A good young pen dog, Toklat, came in and played with her. Mothered by a hysterical little pen dog, Bonnie, he had been fathered by Baranof and would grow into a big black and cream-ruffed "sled dog" of weighty character.

But the one that restored Wildie was cocky little Timmy. Determinedly he played with her whenever she was in the courtyard. She went down in the course of a romp once—those hindquarters had failed her—and for this once she lay still. Timmy tried to stir her to play. He seized one of her hind legs; he held on. Still she did not get up and play. His eyes lifted for a moment from the leg to her face: Aren't you going to respond? Then she did, she played.

Her spirit was not affected. She tried everything and did not get discouraged. But when her legs would not do what she wanted, she whimpered to herself in her odd little baritone voice.

One day, five weeks after she had been carried helpless from the pen, she walked steadily up to me among the dogs and kissed my hand for the first time. A few days later I felt her warm, firm, searching tongue on my eyelids—Wildie kissing me as I laid my head under hers on the couch.

I trusted her, she was able, and to her delight I began taking her along on the daily dog walks.

In all we lost thirteen animals from distemper, including Bobby and Wallie, who seemed to recover but later relapsed. Alatna lived. If she had died now I should never have known her, nor realized that I did not know her.

15 *Wolf-Dogs*

IN DECEMBER the unfinished theme of wolf-dog nature began playing swiftly to a conclusion.

Status changed in the big pen one day. There was a fight, over in a minute but real, between Kuskokwim and Kotzebue. Between Katmai and Alatna. Also Katmai growled Alatna away from a pan of food. She had never done that before. (At the same time she gave a placating wag of her tail!) I hated to see the status of the two leaders challenged.

Katmai still "swept a curtsy" to Alatna sometimes. This is a striking wolf gesture of submission. I had witnessed the young wolves in the Arctic perform it toward the dog Tootch. William Berry, the animal observer and artist, witnessed and with instant pencil recorded it performed by one of a band of wild wolves he was watching. What happens is that, with no preliminaries that I ever noticed, a younger wolf steps up to face an elder and all in one fluent motion both bows her head, ears erect, and sweeps her bushy tail slightly aside. The flowing beauty of the wolf tail and the grace of the gesture make the term "sweeping a curtsy" seem inevitable. Alatna received Katmai's curtsies impassively but her eyes looked stern.

Kuskokwim and Katmai changed toward me. Until now they

had been encouragingly friendly. They had taken up the game of jump with me. Kuskokwim had even let me stroke his back. The first time this happened he had frozen as if in ecstasy, leaving his legs scattered awkwardly where they stood. Now I feared him and Katmai. She would try to slip around behind me in playing jump; I distrusted her. Kuskokwim shrank from my hand and quailed in the game as if I were hostile.

"He *knows* it's only a game," I thought once and jumped impatiently toward him. He gave me a dull look and withdrew. More harm had been done. However, the inexplicable change in him had already occurred.

Kotzebue alone of Alatna's first litter had not changed. He was still as always aloof, shy and friendly. I never feared wolf-heart Kotzebue.

I no longer dared to go out into the big pen. If I had to open or close the gate to it, I did so only when the hostile pair were absent. No longer could I please Alatna by appearing unexpectedly in her haunts. I could not even play freely with her: I had to be on my guard against Kuskokwim and Katmai.

I took a chance one day near the end of December: I went down and opened the gate, letting them into the yard. If I had come up to the cabin when they made their first dash up to the plaza—to look at the pen dog Bonnie and her new puppies in a closed travel box—all would have been well enough. But Alatna, pleased to get out of the big pen along with them, wanted to stand up and kiss me. I was so pleased that she still wanted to do this, in spite of my recent reserve, that I waited and let her kiss.

The two came around while she stood up but they did not molest me. But when I moved away from her they started to chivy me, Kuskokwim in front, Katmai trying to keep back of me. Her eyes were sparkling, his were cold and lightless. If I had stumbled as I backed up the steep hard bank onto the plaza with the pair baiting me, it would have gone ill with me. After I was on the plaza I tore off my jacket and swung it a little in front of them as I backed toward the cabin. It was well that I had not done this sooner. It made Kuskokwim very angry. He foamed up all of his fur and became more dangerous. I got to the door, backing.

Meanwhile Alatna's look was unforgettable as she watched me backing, them tormenting me, not qui-ite venturing to close in. Her look was not possible to interpret. It seemed to me surprise and pity. (I was afraid I had so lost prestige in her eyes by being backed, put to flight almost, that I would lose face with her for good.) Only one thing was clear—her total alertness, her ears, her clear eyes intent.

At dawn, two days before the end of this troubled year, the maverick pair were executed. We planned carefully, foreseeing how they would act.

When all the pen animals were on the plaza before breakfast, Cris ostentatiously went on the outside of the fences to the drive-in gate down the avenue. As expected, Kuskokwim and Katmai raced off to challenge him. As soon as they were out of my sight I ran down and closed the gate into the big pen. Unfortunately Kotzebue slipped in and had to be shut in there too.

Alatna was terribly frightened at the first of the two quick shots. She fled to the home pen and I shut her in there, with Baranof for consolation. I was not associated with the killing in her mind. Cris came for my help and together we pulled the bodies up over the palisade into the garden.

It was no use keeping Alatna out of the big pen. Four hours later and it would still have smelled the same to her. I let her in at once. She did not howl but smelled the spots, especially where the two had been taken up over the palisade.

Kotzebue suffered the most. He looked and gazed and watched for the two. They were his siblings. He howled.

He did one extraordinary thing. Aloof, shy Kotzebue came directly to me on the plaza and howled and also "talked," looking into my eyes. He was appealing to me for help in his terrible trouble. I should have walked along with him as if to help in his search; I should have helped him somehow to live beyond these minutes. But absorbed in my own stress of feeling I did not remember that he, like Alatna, did not connect me with the disappearance of his fellows.

"Yes, it's bad, Kotzebue," I admitted truthfully. At this he went away—would have done so anyway no doubt. He grieved.

I tried to distract him and Alatna. I gave eggs. I whipped cream and gave that. I gave Kotzebue the slice of fried ham intended for my breakfast. I gave Alatna a box of dog candy to tear and share. I brought the Braunschweiger sausage and gave thick slices to all, especially Kotzebue. What terminated his and Alatna's to-and-fro-ness was perfume. I emptied a Christmas gift bottle of perfume on the deck and a travel box. The two were tired by then and soon went to sleep.

The rest of that day was blissful. No "hating" through the fence of the dogs' courtyard. Katmai had hated Wallie—whom she had exiled—and yearned to kill her. Kuskokwim had hated Timmy, yearned to kill him. The day was quiet and peaceful. No endless unhappy howls from Kuskokwim. For the past week that wolf-dog had not been happy. I walked through the big pen without fear. I had felt no grief as I looked at the beautiful bodies. Those two animals were to be pitied.

Alatna was so gay and calm all the rest of the day of the execution that I realized from that how much tension she must have felt, trying to keep her supremacy with those two. I remembered how she would gnash her teeth and fly at the weakest, Mr. Shy, to intimidate the strongest, Kuskokwim. Now she frisked and played jump with me. (My own relief no doubt lightened her spirits, too.)

And I had played jump with Kuskokwim and Katmai. A week before this I had noticed that they changed and were more dangerous. They had not looked at my eyes, only at my legs, gnashing their teeth, drawing back their lips, looking delighted. I had got onto the deck with my back to the door.

"I'll not play jump—ever—with Doonerak or Shy," I resolved. Already Doonerak had cold eyes like Kuskokwim.

But the game alone had not caused the change in those two. Katmai had long since turned against Bobby, later against Wallie and then Wildie, at last against me. Had her intolerance been due in part to overcrowding? A new concept dawned on me, that of "psychological carrying capacity." Perhaps our pens, big and varied as they were, had a psychological carrying capacity as well as a physical one. We would not exceed it again. We would not let population build up to the point of psychological stress. It would be a hard resolve to keep. The alternative would be harder.

I still felt that my question was not completely answered: Why had those two turned against me? I wrote the painful question to Carol, our psychologist friend.

"Now, Lois," came her reply, "my offhand, ill-considered but perhaps not inaccurate observation about the wolf-dogs. You will bridle at either my utter determinism or my mysticism or both. There is an accumulating document of evidence that schizophrenia has some strong hereditary components. I don't know. But mix a strain of a species that owes its very existence to developing behavioral characteristics of subservience—the hand-biter was exterminated eons ago—with a strain that existed because it was group or pack and had wolf-oriented subservience, behavioral characteristics fostering pack survival, and one has, I think, a condition of irreconcilable conflict.

"The wolf-dog cannot eschew the pack. Pack he must have, but how intolerable his pack! Perhaps in turning he repudiates both packs, the human and the wolf, so he has neither and needs both! This, Lois, as I understand that incalculable anguish of schizophrenia, is what happens. The urgency to be wolfish and doggish and the total impossibility of being either or of being both.

"All I am trying to say is that the wolf-dog could well be born with built-in conflicts calculated to maintain the creature in an almost chronic state of anxiety or rage. At any event, their behavior might well be unpredictable because it is prompted not by the past and present conditions of reality but by an inner condition of tensions or needs which develop or occur, heaven knows how."

I continued daily life with wolf-dogs, trusting Wildie and

Kotzebue without reserve, sure already that Doonerak was hostile.

Wildie's last day of freedom for walks came in February. On the walk that day she came homeward gaily with the dogs and me to the foot of the draw. There she stopped, watched us start up the road, then turned decisively and walked away. She returned later, but Cris reported that evening that her tracks had been seen clear over by Enid's cabin, two miles away. That settled it. "If she goes that far," said Cris, "she's liable to chase cattle and get into trouble."

The next morning I "loved" her, taking more time than I usually spent. She kissed me and loved me back. At walk time I managed to get her into the big pen, her first entry into it since she had been carried out of it in September. While the joyous reunion with her wolf family was going on, I slipped away with the dogs.

She was still radiantly happy when I came home. Her ears were wrapped around her scalp. She was following the others everywhere and courting them. But she was black-wet from being knocked down many times. (Her hips would never be steady.) Now she wanted to return to the courtyard, from which walks started. I got Alatna and Kotzebue shut into the big pen and let Wildie into the courtyard, along with Shy and Doonerak and the little pen dog Bonnie.

Bonnie knew the courtyard: her puppies had been born there, in the woodshed, a precaution against Alatna, who could no longer be trusted with birth-smelling dog puppies. Bonnie's puppies were old enough to be safe now; they were ensconced in the walled-in hay nest under the deck table.

Doonerak and Shy were interested in the novel environment. Wildie was bored. She knew what she wanted: Out! She studied the big gate at the far end of the garden pen and looked away through it. Such ideas would not have occurred to the inexperienced animals, Shy and Doonerak. I lured them back into the yard pen at last by dragging and shaking a rope along the ground. They followed it curiously. Wildie stayed behind, lying in one of her old beds down in the garden. (From this time on, she and she alone would have the run of both the dog and wolf pens.)

Next, while Bonnie was still happily running in the garden, Doonerak fulfilled what must have been a deep desire. I re-entered the yard pen to find Shy watching the puppies' nest-den. Inside it, curled up with the puppies, lay Doonerak.

Bonnie returned. Barking indignantly, she got into the den, separated by Doonerak from her puppies. I ran to bring a camera and heard all hell break loose on the deck. Bonnie and Doonerak were fighting inside the den. The puppies were screeching. Alatna came bounding up and tackled the guilty one, Doonerak, and put him down the bank. Then she jumped on Bonnie and pinned her down and lightly bit her muzzle. The minute she was released the unwise dog stood up on her hind legs, loving Alatna's muzzle.

Meanwhile the scared puppies huddled at the door, wanting into the cabin. I took them indoors. The next time I opened the door, a wolf present lay beside it, a chunk of beef heart that Doonerak had dug up somewhere and brought to entice the puppies back outdoors.

"He's going to be Father Wolf," we agreed that evening. Doonerak had two marked characteristics both of which would make him a great father wolf. He had an aggressiveness the other animals feared: they yielded to him, he got the food he wanted. (Once he cornered a whole calf carcass.) And he had this overpowering zeal to be with baby animals, to cater to them, play with them, have their paws pat his nose; above all, to bring them food.

Alatna entered her last pregnancy by Baranof. The big dog was old and ailing. During the height of estrus she was luxuriously kissing in mood. It is an experience of beauty to be kissed by a wolf. The flash of fangs and the clear orbs are inches from one's own eyes. Alatna slept with completeness during the early days of pregnancy. Her eyelids formed a calm closed line as she lay in the sun while I stirred about on the plaza.

"Sex means more to wolves," I thought, "at least to this female wolf, than it does to dogs. It is not transitory but is a kind of continuing happiness—to have a mate near—and a cause of concern if the mate is absent."

I looked around one March morning to see how many animals were in the cabin as I prepared breakfast. Seven dogs and a wolf. Four of the dogs were puppies, two each from two mothers. The littlest gray pups, Spotty and Curly, hung to the hem of my jeans, tugging and growling as I waded around. Alatna also waded among the dogs, her tail happily elevated. She kissed briefly each pup and dog. She lay down on the bunk and through the open door of the bunk nook watched me attentively. Now and then I gave her a tidbit of butter or crusty brown ham fat.

Meanwhile Cris dragged a bale of hay down to the shelter under the big spruces and clipped the wires, leaving the wolf-dogs to open and distribute the hay for their beds. The sound of happy, absorbed growl-crooning came up to me as they wrestled and busied themselves. Out the window I saw Wildie intently looking up, a spring checked in her muscles, at a groomed Clark's nutcracker tugging at meat nailed to the top of a fence post. Sunlight came over the mountaintop, edging the trees with light.

Mornings like this at Crag cabin were dear to me—the sweet high freshness, the morning light descending from the sky-crags above, the bright-eyed eager animals.

Alatna denned one evening in late April, earlier than in other years. Lying in her den the next morning she flicked her ears down as she saw me—that lovely "glad to see you" of the wolf. She had not given birth yet.

I drove to the Lake and supposed of course she would have had her puppies when I returned, but no. I heard a growling commotion and was surprised to see her up in the yard, thronging with the wolf-dogs, growling harmlessly and riding Mr. Shy. She seized a shred of a torn red bandanna, a bright arc hanging from her mouth. She looked misleadingly gala with that rag and above it her brilliant eyes. She dropped the rag carelessly. She was panting hard.

I rushed the dogs she hated out of the cabin, Timmy and a few others, then opened the door. She hurried in, tore up the bunk, was out and in several times, coming urgently, head lowered, plowing straight toward the bunk as if there she might find ease. She panted hard and quiveringly. But she turned impatiently from a proffered pan of water. She snatched a woolen shirt, used as a puppy bed, and dragged it clear down into her den, brooking no interference on the way. (For once her den would be lined!) She returned and I was afraid would deliver a pup on the bunk, which would have made a problem about getting it home to the den. But she hurried in a crouching way down to the home pen, where she pounced again on Shy.

Amusingly, Kotzebue was nowhere to be seen. He was strictly absenting himself. He had suffered Alatna's prebirth antics the year before and once was enough: he did not propose to undergo them again. This year harassed Mr. Shy was bearing the brunt. Alatna followed him around the home pen and mounted him or seized his back in her jaws. He looked disheveled and subdued and submissive, and as if he would like to avoid her. The whole retinue of animals, except for Kotzebue, made quite a throng around the den mouth.

Alatna hesitated, went into her den. There was a hoarse "Ow. R-oo," and a yelp. In a moment came a baby-puppy cry.

A day or two later I was chopping fresh beef heads at the chopping block in the dog courtyard. Doonerak lay inside the yard-pen fence, watching. So I gave him a meaty chunk and for once he held it high away from the beseeching young pen dog Toklat.

"That's odd," I thought, "He doesn't usually coquette."

And this is what he did. He hurried down to the den mouth and coaxingly gave the wolf puppy-call. He wanted to present that hunk of head to the little newborn blind puppies. So, to give him some gratification for his father impulse, I shut all the other animals into the big pen and, forgetting two facts, let the little gray dog pups Spotty and Curly into the yard for him to "love." He was beside himself with delight. I had forgotten that the gate to the home pen was open, and that Spotty and Curly often fought.

They fought. I knew where I was in an instant. It was life or death all around. Doonerak menaced me. All the same I seized Curly and got her into the cabin. Alatna came raving up into the yard, all foamed up along her spine. Spotty, the dear jumpy little puppy, lay between her paws. She went for his neck—fiercer now because she had puppies of her own. There were no screams. A quick kill. So I hoped.

She carried him hanging from her jaws down to the home pen. I saw his little paws hanging motionless as she pawed and mouthed him. I turned away. There was only one thing to do—try to live beyond this horror quick. (I *liked* little Spotty.) To dwell on it was useless. But as I tore into the closet shelves to straighten them in a kind of desperation, there came the harsh cries of agony. Spotty was not killed yet.

I screamed. In solitude one can do that. I fled from the place with what wrap I could snatch, ran to my car—I was thankful for it—and drove to Lake George. I talked with friendly people, saying nothing to them of the trouble at home. That evening I took a sleeping pill and slept.

But first an odd thing happened. Down in the dark, from the direction of the den, came a wolf puppy-call—Alatna's, I thought. "Calling the blind puppies to the feast," I thought with bitter irrationality. I thought I would not feed her again in the den. I could not bear to see the mangled little body.

But in the still gray shadow early the next morning I did take down two pans for her—warm milk and whipped egg yolks. As I neared the den a strange little figure came over the rocks toward me. It was Spotty. I felt no emotion. He walked stiffly and looked somber. His neck was mouthed. I set down the pans swiftly and silently where I stood, and lifted him. I expected Alatna on my back if we made a sound. Swiftly I walked from the pen, fending off Doonerak. Not daring to take an instant to glance behind I got into the cabin with Spotty.

He had a fine tremor, a tetany. That had been his first night in the open. I warmed milk; he drank it in my arms. Then a piece of frozen beef heart went into the oven, a bed was prepared on the couch. I knew how one may feel after a bad rock climb: one wakens with a jerk if one relaxes and his body slips. So I supported the puppy on three sides with warmed pillows and an old wool shirt. He was so dazed that he urinated on the shirt. He slept, still with an old face. I cut warm juicy heart and wakened him for that, then he slept again. His neck was swollen but the skin was unbroken.

Forgivingly I took the rest of the heart to Alatna, in the den.

A couple of days later a dog fight broke out in the back courtyard. Alatna, shut into the home pen, was leaping along the fence, foamed up, crazy to get to where she could view the fight. Knowing how she felt, I ran down and opened the gate. She checked herself as she ran past me, and kissed my hand—a mere flick of the tongue but so responsive: I had opened the gate for her.

16 *Last Puppies by Baranof*

THE TARRYALL valley and regions around were a country of lonely ranchwomen. On a rare visit to a widow living, like myself, up a long road behind closed gates, I was showed around her warm clean house. Lengthwise on a bed lay a rifle, the husband of fear. A young ranch wife whose husband was away working told me two drunken men were terrorizing outlying dwellings in the county at night. "I have a gun and I'll use it," she assured me resolutely. Waiting at an isolated ranch for cull ewes to be singled out and shot for me, I heard the ranch wife beg her husband tearfully on what was evidently an old issue—that they move away to a place among people.

An old-timer said the Tarryall valley never had been sociable country, even in his grandfather's day when wild grass was deep, buffalo and antelope existed and one might have expected pioneer warmth to prevail. Now some of the few ranches had been sold to outsiders for the fishing rights; former owners moved away. A local telephone line that ran the length of the valley fell into disrepair and was abandoned. There was no church within scores of miles, no community life in which to join. All this made a kind of aridity and harshness in the valley.

I was apart from it, seldom lonely, uniquely situated to keep my "pact" with Alatna. (The pact did not haunt me; it lay deep in my mind and will, thought of chiefly in May, the month when our wolves had left.) I had the companionship of big devoted animals; it was not to be underrated. Alatna was an equal companion, full of surprises. The outside dogs with their differing personalities were companions.

Beauty, too, gave companionship. When blue smoke rose backlighted into morning sunlight from our dark log cabin among shadowed firs and mountainsides I was tantalized again and again to try to capture with a camera the winsomeness of this human dwelling in near-wilderness beauty.

I tried to gentle two of Alatna's new litter for companions. I removed them from the wolf pens. But they remained too close to Alatna: she could hear their disconsolate howling and grew feverish. I had to return them to the pens.

For the first time females predominated in the litter, three to two: Haida, Yukon and, wildest of the lot, Black Beauty, not black but dark and a beauty. Of the two males one was placed; remaining to us was Noluk.

Alatna was a widow. Baranof had been euthanized by the vet, who could not save the big old dog's life. Alatna did not seem to miss her nonmate: he had never been Father Wolf. Being widowed would change her profoundly but for now the change was deferred; one more year of happy matriarchy lay ahead for her, with both young and old offspring around her. Kotzebue was a mate-surrogate, though there was no sex play between them and he never, never encroached on her status. He was the eldest wolf-dog and the only survivor of her first litter except for unknown Kobuk, down at Colorado Springs.

No escape had occurred for a long time, not since the repair of the fences. But late one afternoon some of the wolf-dogs went out. Notified by the dogs, I arrived on the scene a minute later. Wildie and Kotzie were out. So were three baby pups. Alatna was frantic. She was watching the direction they had taken. I ran the other way, up along the back fence, to find and block the escape place. I found Shy digging hard at the spot, though he could have squeezed through. I scared him away and dragged up a tree trunk and rammed one end of it into the dig. I ran and brought other heavy objects to block it. I got Alatna and Shy confined in the home pen. Later the left-behind puppy, Black Beauty, came over and all three were united. But that did not happen until a couple of hours later.

Meanwhile I stayed in the home pen with Alatna. I thought she would feel that I shared the anxiety and tension. She was terribly disturbed, panting so hard, her tongue hanging so far, that I got water to her and Shy. This was not easy to do because she was determined to go through the gate if it was opened. At first she kept leaping straight upward, her tongue swaying. Then, when there were no more stimuli of sight or hearing or smell from the free wolf-dogs, she did not leap any more. But she was not at rest for one moment, except in a touching way as I will tell. Back and forth she hurried to different places along the fence and started digs. Most impressive of all was to watch her considering where to dig.

If I say "started digs," that implies numerous random digs. But they were not random. She made a quick approach to a possible place, gave a black-eyed glance at the ground and to either side of the spot, and a swift appraising upward look at the fence. All was done in a moment and decision made—no good here. But at last a point would be chosen. Then, digging, she encountered the buried fencing. The pen held. But I did not think it would hold.

I was exhausted, partly from my feelings at the thought of what the free wolf-dogs might do to cattle and of their own fate. I did not doubt that with two big wolf-dogs and three puppies together they would be gone for good. Partly I was weary because of going constantly to watch Alatna's new efforts. It was to block them—how I did not know—if they threatened to succeed that I was staying in the pen.

The touching pause was this. Alatna would come and stand against me, and my "loving" of her seemed to ease her distraction for a few minutes. She came again and again to be stroked and hugged. Then she hurried on. But that seeking—and accepting—of a moment's pause in her troubles was a moving thing.

Shy did not take the situation so hard. When the free animals were clear gone he even tried to play with Alatna. It was of some help in her distress to have him paw her. And when Black Beauty got into the pen he enjoyed playing with her. "Shy is a nonentity," I thought severely. "He has no character." Black Beauty felt worse than he did. She fretted and hunted for her lost siblings, and whimpered and howled.

When the air chilled at sundown I managed to slip out once more and return with a wrap. Cris was due to be late this evening and he was. But unmistakable at last, above the rush of waterfalls in the ravine back of the pen, came the sound of the engine, then the lights of the truck. Help coming.

That night was short and disturbed for all of us. We worked until midnight, trying to get the baby puppies into the pen. They were in the woods on the mountainside. They howled. Stumbling by flashlight among crags and trees, we could not find nor shine them. They fled.

We lay down and slept but Cris got up at 2:30 and soon got me up. He had the puppies in the pen, had herded them through the meshes of the fence. Even with her flock of babies all together again, Alatna could not rest. She paced and hurried, with them all following, a retinue, wanting to take them away into the woods of the big pen. We could not let them in until daylight, when we would check to be sure my desperate blockade of the escape dig would hold.

Kotzebue was in the garden pen. Cris went down and closed the big gate at the far end; I had propped it open in case some animal did return. We lay down again but when the light grayed we got up. The garden pen would not hold Kotzebue long.

From here on all was easy in one way, heartbreaking in another. Pups and mother and Kotzebue were all at home in the big pen. And poor pretty Wildie, whom I had loved, lay curled just outside the fence, wanting in but not having connected with the gate we had at last been able to open for her. She had come to the gate before the dogs were secured so that it could be opened, and had gone on around the pens to where only the fence separated her from her family. I could not call her back once the gate was open; she did not understand a call. Cris shot her. It would appease Enid. But the main reason for the act was the deadly thought of overpopulation; four new young wolf-dogs were in the pens.

July held one more emergency. Alatna's crying called me outdoors one day and when I saw her plight I did not dare move for fear of making it worse. She and her four scrambling puppies were on top of the fence tower against the sky. The puppies had evidently just discovered the place and were trying to go through the wide-meshed, loosely held wire at the edge of the fearsome drop-off—forty feet or so—into the steep ravine.

Alatna was terribly agitated. She put her jaws over one puppy, trying to walk it away from danger. She struck her foreleg like an arm over another that was starting to crawl through the fence, curled her paw strongly and hauled the pup safely back toward her—an impressive new action.

In late July Black Beauty was given to the master of Kobuk. He had married and now lived in the Black Forest near Colorado Springs. Neglected, Kobuk was desperately lonely. Black Beauty, three months old, was too old and too wild of nature ever to become a pet but she would be a companion for Kobuk and in time, it was hoped, his mate. The female wolf-dogs came in heat and mated earlier than pure wolves.

August was a splendid month. Alatna relaxed from special maternal cares and became serenely the generalized matriarch. It was beautiful to see her and Kotzebue striding up the lightly shaded avenue side by side, turning their bright glances toward each other. Following them came the other animals in line. Bonnie, the pen dog, ran offside from the two chieftains, looking toward them and barking hysterically with excitement. Bonnie played the fife.

I had intended to sleep late one morning but persistent yipping from the pens got me up about six. The pen dogs were barking "treed" and for fear they had a porcupine I started over toward the bench.

As I neared the deep-shaded brook here came all the animals streaming down the hill to meet me. Alatna hurled herself up to kiss my face, whimpering with excitement. Her eyes were like a gray and sunny sea. Then the animals rushed back up the trail. At the top of it, on a limb thirty-five feet up in a big fir tree, stood a cougar or mountain lion.

He was a clean beautiful animal. He looked understandably bitter and miserable. If he came down something was going to get hurt—the cougar surely, the wolf-dogs perhaps. Probably he had been treed for quite a while. Alatna and her puppies soon went off to their usual lie-up and slept. Kotzebue, Shy and the dogs lingered under the tree. The cougar rested himself by putting his forepaws down onto the limb below. Both limbs were small. One comfort the cat mercifully had: clouds came over and a few drops of rain fell. Besides, the tree trunk gave him some shade. I tried in vain to coax the animals out of the pen so he could come down.

Not until midafternoon did the animals leave him and come over to me at the cabin. Thankfully I ran and shut the gate into the big pen. But my troubles were not over. The cougar's, I thought, were: he could come down in safety. I tricked Alatna and all the rest into the home pen, except timid Kotzebue. For a troubled hour he was caught in the yard, trying to dig back into the big pen and also longing to get into the home pen with the rest. I had to keep blocking his digs. At last somehow I got him into the home pen.

But now the cougar, far from coming down, had relaxed. He hung like Elsa along a limb, legs dangling on both sides, head on one side, big tail hanging on the other. Cris came home at twilight and I took him over to see what the big pen held. He tossed pebbles near the cougar, hoping to make him come down. A pebble hit the cougar's face and he hissed, a mouthful of hissing. He threatened to climb higher and at that we hurried away. He had problems enough to get down from where he already was.

I slipped back with a chunk of fresh beef liver and laid it as a lure on a rock under the tree. I was concerned by something Cris had said—that a beleaguered cougar might stay up a tree for a day or more.

I sat silent during dinner. I did not want to violate some kind of inner intensity that stayed with the cougar out there in the tree, as if to remove my attention might weaken his will to escape. I was not giving thought to the cougar, not imagining him. The feeling was a bond of itself—sui generis—going on alive, intent, underneath words or thought. It was an intense, secret, absorbed feeling.

At dawn the tree was empty. The cougar had come down, taken his liver and left. There was no one in all the countryside we could share our pleasure with; people would have sided against the big, scarce wilderness cat.

The wolf animals were disciplined. They were "law-abiding," but people were sometimes anarchic toward them, having no category which would include wild and semiwild animals.

A woman was at the cabin one October day with the oddest of attitudes. A group was with her—about the last of our usual stray summer visitors. The woman had a movie camera and tried for pictures of the elusive wolf. The beautiful shy wolf-dogs venturing gently up were nothing to her except as tools. She overrode them not only carelessly but unconsciously. When from the cabin and group of people she started down into the big pen, I asked her not to go beyond the shelter. Thinking her safety was at stake, her companions repeated to her what I had said.

But what I thought of was her ruthlessness toward the animals. That they were loved citizens here and had rights, that we had to live with them and did not want them frightened and made wild did not enter her head.

The animals disciplined each other. Also they accepted discipline; they were self-controlled. Once Shy lay on a travel box, that ever-gratifying elevation, looking composedly away as I fed Alatna and the pups on the plaza. But raindrops of saliva kept falling from his jowls.

Haida obeyed wolf "law" one evening. We had given Alatna a rabbit. Haida tried for it. She followed Alatna, who cried protest through her mouth-grip on the rabbit and went at a dead gallop down over the terrace to her special lie-up in front of the shelter. Here Alatna resorted to law. She laid down the rabbit and Haida conceded. She lay two feet from the rabbit; she rolled beseechingly onto her back. But she did not touch Alatna's meat.

Shy seemed anarchic on one taxing occasion but in reality he and all the animals were obeying "law." A highway-killed beaver was placed in the big pen one evening. I supposed the next morning that I would not have to give breakfast, the animals would be fed. But when I went down to say good morning, Shy had the beaver, guarding it, lying so close he could lick its fur possessively after being challenged. It was torn a little at the neck only.

Alatna lay curled glumly in her usual place before the shelter. Substitute objects were in active circulation. Deer antlers had been carried up to the yard. Noluk lay gnawing an old bone. The whole thought of every animal seemed centered on that beaver they could not get. They could not—simply could not—gang up on Shy. Wolf law forbade. The beaver was a unit, Shy had got the unit. It was his property and they respected his rights.

Heartened because I went near him, Alatna went over, held her head low to the beaver in spite of Shy's snatching at her—and wagging his tail deferentially at the same time! Alatna's eyes were impressive as she held her nose down to the beaver and then, imperceptibly, was found to have taken hold of it. Big-pupiled, menacing, they were focused but not where one would have expected them to be focused. She was looking fiercely aside, not at Shy or the beaver. She tugged with Shy over the beaver for a minute, then was bested and returned to her place and lay down, glum again.

Haida ventured. She lay on her back, her soft white belly up, at the unguarded end of the beaver but did not touch it. Then she, too, bit and had a tug with Shy. He was more tolerant of her—and not the least bit deferential, no tail-wagging. I thought she was going to be allowed to stay and eat. But he won. Again he lay curled. Without a move of the head, his cheerless eyes turned under the eyelids toward any threatened approach.

Then I took a picture and for once the flashbulb exploded like a shot. All near animals fled, including Shy. Alatna promptly came over and got the beaver, and Shy plucked up heart—after the scare—to fight her. They fought fiercely, standing up, bowing their supple bodies. It looked terrible but was over in seconds.

Then Shy relaxed and was cheerful. He must have guarded that thing all night. Now he stretched, he urinated, his eyes looked cheerful and bright and he sallied around. Alatna and undiscourageable Haida ate, Haida easily, at the red neck, Alatna yanking with difficulty to break the skin on the rump.

I sensed, these days, that tensions, rearrangements were taking place in the wolf pens. I thought they were due to the fact that Alatna approached estrus and lacked a mate. One of the two males, Kotzebue or Shy, would win, so I assumed. Would one be killed?

The wolf and wolf-dogs were gay in the first storm of winter —wet heavy snow in dark light, a fence-threatening storm. I worked in the fresh wet ozone-smelling air to save the fences, moving along in a "room" softly walled and low-ceiled with fog. I knocked snow off the fences and the bowed trees, jarred it in showers from tall trees that could fall across fences, lifted wet black snags and branches off the sagging wire. Looking dark in the snow, the wolf-dogs ran to me now and then. I stopped and jollied them, then they ran jubilantly away and disappeared among the evergreens.

The work was good, the kind that uses new muscles and leaves one deeply relaxed, ready to rest and be stronger. But then came drudgery with stoves. In the dim gray light of the greenhouse I fired up the balky stoves, trying to thaw the blanket of snow from the plastic-glass roof. At last the snow slid. Then wooden roofs were to be shoveled.

The second storm was different. It was powder snow. (Only the first storm was wet in this climate.) In the sun the next morning the wolf animals were wild with gaiety. I was out on the plaza for some reason, when, thump, thump, Alatna hit my shoulder blades with her forepaws. I turned and the lovely wolf bowed. Her sea-gray eyes sparkled. She tossed her head. The thumps invited me to dance along with the wolf-dogs.

Obligingly I hoisted a booted foot. Generously she took the movement as dance and, wild with responsive joy, she curved her body, tossed her head and leaped around me, while all the wolf-dogs wagged and thronged around with sparkling eyes.

Kotzebue was killed in late November.

He had seen a boundless land. He was determined to lead the other animals out of the pen into it. I had never seen him so purposeful. He was whole. He had always been dominant by age. Shy, in a ceremony of submission, had courted him. Kotzebue would then bite Shy lightly and the latter would scream beseechingly. And Kotzie's eyes would be malcontent and dark at these times. But he was not mean. He was very shy, timidly wary of me. He was, perhaps, the most beautiful animal I had ever seen—very fair, with clear dark dog eyes, unforgettable, like those of a human being looking at you, because of the being, the wolf-dog being, within.

The spark of health was in those eyes, but not of happiness—of gaiety sometimes. Always Kotzebue had been troubled. He was a dual being. He had big dog legs and a tundra heart—wolf heart and intelligence and orientation in a part-dog body. He had not been shy, not troubled that day after he got back into the pen with the others, his pack.

What had happened was that two electric power workers came on a special survey that involved one man's climbing the high pole on the hillside overlooking the pens. The wolf animals were frantic, as they had been when the outdoor club came. That visit had traumatized them; this man up the pole seemed to reinstate the feeling. They were helplessly exposed. Besides, the dogs were in an uproar. Kotzebue, shyest animal of all, must have hurled himself up a fence, climbed out like a cougar. I saw him going away up the hillside and ran to find the escape place and guard the other animals from escape and death.

I felt desperate as I ran through the gate into the big pen. If I went to the right, down the avenue, the animals might at that very moment be escaping up on the bench, to my left, and vice versa. I ran down the avenue. I returned along the bench, ran to the home pen and searched it, then tried the avenue again. I could not find the escape place.

After the intruders left, Kotzebue circled the pen. He wanted to get in. He sat down and howled when he could not get in. When at last he was in, the reunion was beautiful. The other animals thronged his face, kissing him, looking so glad and gay that I was pleased and thought nothing—at first—of the fact that he was leading them.

Without determination or force or anxiety he was one will to lead these animals out of the pen. I had never seen his eyes like this before. They were untroubled. He streamed, tail up level, down the avenue, up and around the hill and bench and back to the home pen. All the others streamed along single file, following him. It was a beautiful sight. I was deathly tired. I kept going along, to prevent escape if it started.

Kotzebue stopped once and looked up at the top of the fence where it was low—the escape place, had I known it. On the plaza he looked away at the hill going to freedom, where he had climbed on first getting out. His eyes were clear, intelligent. No "determination" was in them, just invincible gatheredness, to lead free. The other animals were total followers. Glad, confident, they veered and followed wherever he led.

Alatna fell out and lay by the shelter, watching whenever they passed. I fell out, exhausted, and crouched despairingly on the edge of the deck, knowing that at any moment Kotzebue might "remember" the escape place and lead the animals out to freedom and death. It was his life or theirs.

There must have been desperation in his escape leap. There was no desperation about his eyes or bearing now. He bore himself free. This was the thing beyond happiness. It was authority, purpose, sureness. Kotzebue was himself at last after the clouded years. He was whole, a tundra heart.

His eyes were perfectly friendly whenever he passed me. Willingly he took from my hand the bait holding half of a lethal amount of sleeping pills. He hurried on. The next handful of tablets would be death. I waited. He did not slow down. He was bound to remember the escape place. I held out the other handful.

Cris found his body by flashlight that evening.

17 *Kobuk*

IN THE DARK before dawn on Christmas Day, Cris drove off for the Black Forest near Colorado Springs to redeem the "lost" wolf-dogs, Kobuk and Black Beauty if he could. I did not think he could. Kobuk, after being made a pet and then rejected, had become so fierce that his master no longer dared to enter the pen and had asked us to take back the pair. All day I was in doubt about Cris's success. But I undertook to do one thing not possible to do if he actually brought home the mavericks. At siesta time I started to drag a dead sheep from the big pen toward the home pen, where the newcomers were to be confined. Our animals had five sheep and could spare one.

But suddenly Shy confronted me. He had always cherished property, hated even for me to remove empty pans. So now he menaced me. He made lunges toward me, chest low, gnashing his teeth.

I would have had to drop the sheep, but help came in a lovely way, so sudden and unexpected, like a wink of light, that I hardly knew how to value it. It came from Alatna. She arrived and took in the situation. She rushed not me but Shy and diverted him. She did it in an extraordinary way: she effused radiant gaiety at him like a tangible force. At the same time she was prepared to bite him if necessary. She ran off, with him attending her gaily.

This was a remarkable thing—so remarkable and sudden, so instantly over that I would have to poise it in my hand to value it. For the first time the wolf matriarch had included me under the aegis of her protection. She had graduated from watching intently when I was menaced, to intervening on my behalf as for one of her "pack." As usual she had sided against the aggressor. Six and a half years of tireless affection and concern had gone into producing this moment. It was an awesome little landmark in my relationship with the wolf. But now I had no moment to lose in marveling, Shy might return. I scrambled over the icy snow, dragging my sheep as fast as I could, to get it through the gate and the gate closed.

The first time I looked at the clock it was 1:30. Soon afterward I had such a strong, excited, disturbed feeling that I knew something was happening with Cris. But what? Cris hurt? Wolf-dogs escaped? It felt not good. Time passed. (Later I learned that the caging was accomplished about two o'clock.)

Toward twilight Cris drove up. The cages on the truck were occupied. In Coonie's little wire cage, big Kobuk half crouched, unable to stand; he was savagely angry. Black Beauty was hidden in one of our old wolf travel boxes. Not unloading the cages, Cris fell to work stopping possible escape places in the home pen. Now, before Kobuk was admitted to it, was the only chance to do this. If he escaped, he would endanger people as a pure wolf would not. He had been played with too roughly as a pup and had lost his "awe" of human beings. I held the flashlight patiently, not to scant this work by a single needed nail.

Then the heavy cages were lifted, carried, dragged down to the home pen. There were handle bars on Kobuk's cage but I looked to my fingers more than once, scared by his snarling heavy lunges that displaced weight in the cage.

Cris released timid Black Beauty first, then placed her empty box on Kobuk's cage to prevent his coming back on us. In a minute he too was "free," in the old home pen, where he and his fellows had been born.

The next day I feared to enter the home pen. But after that I could not resist going in. I crouched near the gate, speaking softly and extending my hand. Black Beauty hid–she would always do that or stay aloof. Kobuk merely stood and looked at me. He was a striking animal–a very dark "wolf" with remarkable eyes. He had been through so much, humanized, deserted, half-crazed. Made a pet as a pup, he had been ditched to loneliness and solitude when his master took a bride and moved to an abandoned farm in the Black Forest. Here Kobuk had been confined in a tiny pen and adjacent shed from which he could see only an empty field. The big social animal had nearly gone crazy. It was after we had given Black Beauty to be his companion and future mate that he had mounted into unconquerable ferocity.

I had never seen a "prouder" animal. Pride showed invincibly in his stance, his undrooping tail. His dark head was the picture of pride. Intense pride looked out of his clear strange dark-light eyes. My overtures to him, made daily from now on, seemed to gain nothing.

Of course the arrival of the mavericks was a tremendous emotional upheaval for them, and a big change for our five as well. Our animals hung around the home pen as a cynosure, taking their lie-downs in the spruce shelter close by, instead of up on the bench or plaza. Alatna was benign and friendly through the fence. Black Beauty always, invariably, offered court to all the other females, wagging, even laying herself belly up on her side of the fence.

But Kobuk and Shy, rival chieftains, were all excitement and challenge. They conducted big "hates," watched by the other animals. Bristled and snarling, they rushed to and fro along the fence. They licked saliva off bared fangs without closing their lips, a peculiarly menacing gesture.

The fourth day after his arrival was a low day for Kobuk. From the time I got up until night he did little but lie curled up, indifferent, remote, listless. He came to the fence once or twice but merely turned away, putting out no energy to "hate." He lay on the high table rock motionless, looking downward at me with the rejecting eyes of a captive chieftain. He showed disdain more cutting because so languid: he deliberately turned his back, barely urinated, listlessly executed a faint backward scratch toward me.

Was he sick from the cold? From changing from an altitude of 6,000 feet to nearly 9,000 and from shelter to mere piles of hay in the open? One could see that neither he nor Black Beauty had deep undercoats.

What I really thought was that he was sick at heart. Our animals, after those first days of intense excitement, had begun going off on their own affairs, returning only occasionally. Kobuk understood now that the burst and dazzle of "freedom" —as it must have seemed, after the miserable filthy little pen in the Black Forest—had its limits; that Shy, who had hated with him so absorbedly at first, could run off to interesting places to which Kobuk could not follow and could romp and play with the other deep furs, indifferent to him. Kobuk's wild burst of importance at being inducted into the big varied home pen was over. Disillusioned, he maintained pride by scorning us all. But his heart was so heavy that his appetite failed.

Luckily things did not stand still at this low point. The next day was better, I felt cheered. And on the day after that Cris stayed at home—it was a Sunday—and diverted all the animals vastly by building a shelter for the newcomers. With the inner gate closed against angry Kobuk, he roofed the vestibule of the home pen and in it built two bins, luxuriously hay-filled.

Incidentally, I had put hay in all the good resting places in the home pen before the two were turned in. I had supposed my work was in vain, the hay would be scattered by morning. But meekly they had lain on it, here and there, and never lifted a paw to it. They were more subdued than our animals, who had known more liberty.

It was exactly five days before Kobuk actually lay in one of the hay-filled bins, and another day before Black Beauty lay in the other. On the evening of the day when Kobuk took over a bin I went into the pen with a bowl of tidbits. To my surprise, instead of either eating or going away, Kobuk stood perfectly still, looking into my eyes as I knelt holding the pan. His eyes were clear and wide and light-gold. It dawned on me. He wanted to be petted! Cautiously I stroked his neck and wooled his ears. In a minute he turned to the food and ate cheerfully.

After this I played with him several times a day. I trusted his basic friendliness, though I sensed an almost total recklessness in him. It showed in his black ready eyes searching mine when he was angry. I learned the blunt feel of his teeth on my hand. But I liked and admired him, and was glad that in spite of Kotzebue's death we still had a representative of Alatna's first litter. (Shy represented her second litter and of the third we had four members, Noluk, Haida and Yukon, and wild Black Beauty.)

18 *A New Shy and a Perplexing Alatna*

One of the most impressive changes in man or animal is that from an obscure, submissive underling into a chieftain. Such a change Shy had been making before our eyes.

The Shy-Alatna relationship had always been very subservient and placating on his part. Now he had slid into a leading role because Kotzebue had been eliminated. This was an ambivalent development for Alatna. She had always maintained one attitude toward Shy, namely, to keep him in line, under authority. But now that he stood in Kotzebue's shoes as number one courtier and potential mate, I had not seen her jump domineeringly on him for weeks. She had formerly done so every day or two, pinning him down and making him cry.

And yet, in some indefinable way, she seemed more tranquil. Some hidden tension had vanished with Kotzebue's death. This would remain a complete and surprising mystery to me until events cycled into position to throw light on it.

As for Shy, whereas Kotzebue had trailed Alatna patiently, walking behind her as she progressed through the pens, now Shy did that. Yet with this difference, that he was more easily distracted, more ready to leave Alatna for other occupations—"hating" Kobuk or the male outside dogs through the fences, going off to watch something, a rabbit or squirrel, outside the pen.

His demeanor changed. At almost any time one happened to notice him he would be carrying his tail in happy position—out, down. Also he walked and moved with deliberation, as if his mood impeded his walk. As if, in order to preserve some charmed inner feeling, he had to hold back from panting exertion. (Yet of course he ran bristled, dueling with Kobuk.)

Along with this new slowness of gait, he was also proud-slow-happy about eating. So he was becoming very thin. It was as if his new pride—or that charmed inner feeling—forbade his contesting with mere pups for food. So he stepped around happy-tailed. They ate—Noluk, Yukon, Haida. Did the young chieftain feel the far-off aura of Father Wolf? Or was it merely a way of preserving prestige as the older animal, not to paw and scramble and snatch, muzzle to muzzle in the feeding bowls? Whatever the cause, I had trouble feeding him. He preferred being "proud" to eating.

When he did decide to take from my hand some chunk of meat, a lady could not in a more leisurely way have taken a cooky. He was so deliberate, so delicate with his jaws that the chunk might have been a feather passed. You wondered whether he would make it, receive the meat before eager-eyed Noluk seized it from under his nose. Shy always got it!

His attitude toward me also changed. He had always been cold-eyed; I had distrusted him. So I had begun to court him—kneel and offer my hand though I knew I could not touch him. I offered my voice and eyes, as well as hand, *to him.* He took this in. I avoided ever touching him when he passed unaware. The result of this abstention was that now he did not bother to rise but lay still in the spruce shelter as I passed within a foot, going toward Alatna.

At first I thought he would never change. But he was thawing from within. At last, after some weeks, quite warm and rapid, the thaw began to show. It showed in his changed shining eyes, his glance seeking mine, even, finally, a tendency not to evade my hand. I had stroked his back, barely!

In their own way the young female wolf-dogs, Haida and Yukon, acknowledged Shy as chieftain. They treated him as they had formerly treated Kotzebue, teasing him into noticing them and shrieking beseechingly when he mock-bit their muzzles. Afterward they acted ecstatic. The delight showed an element of ceremony about these aggressions; it eluded my full understanding.

A similar occurrence, not connected with Shy but interesting in itself, was that brash Haida would court Alatna, deliberately provoking her to aggression. Alatna's eyes would bulge to globes. She lunged snarling at Haida, who did sketches of placating acts with all parts of her anatomy. She cowered, she curved her tail under, but the mere tip of it made hints of wags. At the same time she half lifted her right forepaw and licked out her tongue propitiatingly. These little sketch-moves of tongue, paw, tail tip were pretty and amusing. Afterward she would scamper off gleefully, practically on her elbows, throwing her paws sidewise, flattening her ears, as if elated by the attention. I wished I knew how it all looked to her—being jumped on, definitely not being scared though playing her full range of placating actions, and then acting pleased. Her glee puzzled me. It did not lessen my mystification to remember that Alatna, when young, had once acted beside herself with glee when I scolded her, as I almost never did.

Shy's status as chieftain was confirmed by his splendid "hates" with Kobuk. An incident one day made me realize that these, too, were ceremonies.

Kobuk and Black Beauty had a play-dig going that threatened to lead out under the fence into the yard pen. During siesta time one afternoon I began stealthily carrying in rocks and placing them in the point of the dig. The last rock was too heavy for me; I had to ease it down with a thump in the vestibule. Suddenly Kobuk stood before me, black-eyed and growling. To my horror the rock prevented my closing the outer gate. I could only crouch over the rock, my head bent, and protest in a quiet, ordinary voice. I became aware that Shy had arrived back of me. The rival wolf-dogs snarled and threatened with nothing between them but my helpless bent form. I expected that the next minute they would be fighting on top of me and that Kobuk would go out dangerously into the main pens.

In a low commonplace tone I complained peevishly, "You go back. Go on back." To my profound astonishment this they did. Each retired to his own side of the home pen fence, where they began charging fiercely to and fro as usual.

I gave the rock a superhuman shove, closed the gate and was safely outside in the peaceful sunshine, mirthful and incredulous.

On a mild January evening of light overcast muting out the last sunlight, the wolf-dogs played, streaming in a line through the woods. Alatna stood watching. Was she downcast? (This was my constant question.) No; a glimpse of her tail showed it elevated. In a moment, as back they came, she leaped over the nearest wolf-dog and off they all went, Alatna swifter than any others, crippled as she was. Her estrus, the first since the loss of Baranof, seemed to be going well.

But one day in mid-March, when her long estrus had closed, she showed a weariness that was unlike her. She had tried to break up hostilities from Haida, the aggressor as usual, toward Yukon. She had succeeded briefly by riding Haida, who wore the intent grim abstracted look the wolf-dogs had when tackled in earnest by their mother. But the minute Haida was released she returned to Yukon and stood stiffly with her neck across Yukon's neck. Yukon dared not move; to do so would have been the spark in an explosive atmosphere. And Alatna, giving up her peacemaking effort, came over to me with this unaccustomed weariness. I did not connect the weariness with the recent close of her estrus.

We assumed that she was pregnant by Shy, though we had not witnessed mating. But then we never had witnessed it after her virginal first season with Baranof, when mating had occurred openly on the plaza. It worried me that after an interval she bled again. Had she miscarried? The bleeding stopped and I hoped that all might yet go well with her.

I rose one May morning to find her in terrible condition. She cried incessantly. She would not eat. She panted and trembled. Most alarming of all, she crept to the back of an old unused den dig and lay there like a sick dog that has hunted a place to die.

I had a speaking engagement that evening at the state college at Fort Collins, a long drive from home; so I would have to leave her. But I waited until Marie, my new helper, came. Marie was a quiet, sturdy, sunny young farm wife who now came twice a week to restore beautiful order and cleanness to paw-tracked Crag cabin. She could not touch Alatna but she did not fear her. When Marie arrived she promised to comfort Alatna with voice and attention. She would return in the morning to feed the animals and check up on Alatna. Cris was absent.

I drove by way of the Springs in order to consult the vet. He said it seemed Alatna had an infected uterus; if so, the only cure would be an operation, though an antibiotic might help. I knew an operation would not be possible. If Alatna's condition persisted, she would have to be put out of misery and by me: only I could give her death without causing fear. In deep distress I drove on toward Fort Collins.

Arriving late in the afternoon, I was met at the outskirts of town by friends, who piloted me to their home. I told them about Alatna. They put me in touch by telephone with a noted veterinarian, a professor at the college. Like my own vet, he thought Alatna had an infected uterus and that an operation would be necessary. However, terramycin might help; he would telephone a prescription to a drugstore.

The dog Timmy was with me. He was such an agent provocateur toward the other outside dogs that he could not safely be left at home. With silent obstinacy he stayed beside me. At dinner he sat by my chair on the deep-piled blue carpet, eyed distrustfully by my host and hostess, who were not dog people. But Timmy did not seek scraps to soil the carpet; he wanted only to be near me in this strange environment.

My hostess slipped backstage before the lecture to assure me comfortingly that she had the terramycin. At home after the lecture she made preparations for a quick breakfast for me, even to measuring the coffee into the percolator.

But wakening before sunrise I could not linger. I put Timmy out the back door while I dressed—he had slept on his mat by my bed. Suitcase in hand, I stole through the empty kitchen and out into the cool long beams of the first sunlight over the lawns. A policeman came to meet me, to apprehend a dog (undoubtedly Timmy) whose early-morning barking had often disturbed the neighborhood. This would never do! Desperately I told him about Alatna. He apologized and let me and silent Timmy go.

It was noon when I got the first dose of terramycin into crying Alatna. A note left by Marie said she had been crying that morning. From noon on I gave terramycin and aspirin every four hours, rising at midnight and 4 A.M. to keep the medication going.

In the morning Alatna was worse. She was adding a piteous, pleading "Oh" to her cries. Her hips trembled. Her cry trembled. Her eyes looked miserable and distraught. The wolf-dogs smelled her posterior, by no means a common act; it seemed ominous confirmation of the diagnosis.

I endured my life that day and night, hoping the drugs would take effect. They did not. The following morning I drove to the Lake and telephoned to the vet at Fort Collins. Letting the dollars roll I told him Alatna's symptoms, then listened with my heart for his reply.

"It's hard enough," he pondered, "to diagnose for a human being that can tell you where he hurts. It's harder with an animal. But for an animal you haven't even seen! And a wolf at that!"

I begged him to make the overnight trip to Crag cabin just to look at Alatna—he could not touch her. I offered all the financial inducement in my power. Kindly, realizing my distress, he refused: "It would be unethical."

I hung onto the telephone connection, the thread of hope for Alatna's life. Hesitantly I offered one last detail, too trivial for a doctor's notice. "She carries things. She picks up an empty pan and carries it."

Silence. Then a big hearty relieved voice. "Oh! Well, then! It may be pseudocyesis."

"Pseudo what?"

"False pregnancy. Dogs have it. I didn't know a wolf did. Give her a chance. Hang on for a few more days."

With heavenly relief I thanked him. I hung up the receiver, ran to the country store and purchased tablet and envelope. Joyfully I wrote to my gentle hosts at Fort Collins, "My wildie-girl will live!"

As for Alatna's second bleeding, it was probably nature's way of giving her another chance at the mating she must have refused.

19 *Noluk*

THE AGGRESSIONS among the animals were not all of the same kind. Some were open and clear-cut and would have resulted in death if the animals involved had got together. Others were connected with a special place. Some had a puzzling element of ceremony; they gave great pleasure to the participants. Some were mixed, of two kinds at once.

The wolf animals had hate orgies through the fence with the outside dogs. Both parties showed their extreme pleasure in the big hate with waving tails. Two special enemies were Yukon and the dog Tippie. Yukon in a perfect ecstasy of helpless power—she must have felt it in her every strong muscle—actually squealed: "R-r-ee." You could pick out her tiny shrieks above the barks of the dogs and the hoarse wows of the wolf-dogs. She would throw herself down and roll over in a twinkling in order to do *some*thing, if not what she craved. I felt sure she would have killed Tippie if she could have got to her.

At the fence Alatna directed her hate toward Timmy. I think she would have killed him if she could. But she never got her responses mixed up. I could pet her while she snarled at him in a black-eyed rage and she delivered unerringly a lick to me, a snarl to Timmy. I delighted in venturing to test the angry wolf.

(But I did not trust Kobuk to keep responses untangled. He would stand up to play with me and Black Beauty would nip his tail, her eyes snapping with mischief. He growled, he wanted to bite—someone. I was afraid he would snap at my face because Black Beauty was biting his tail. His friendliness I did not doubt, but I knew he was reckless and almost without inhibitions.)

In contrast to the hate orgies at the boundaries of the pen was the friendliness of the wolf animals among themselves. There were the quick signs of affection—the slowly waving tails, the "smiling" faces, the kisses given with a sweet friendly expression of the eyes. Alatna would kiss Shy's face. Mr. Noluk would walk alongside her for a minute and kiss her face.

All the same there was aggression between Yukon and Haida. They had reversed roles: whereas it had formerly been Haida that bristled across Yukon's back, now it was Yukon that bristled across Haida and Haida dared not stir a muscle. I thought that a death-fight would begin between the two at any time. "Perhaps Yukon had better be killed," I thought sadly. If a fight occurred she would win and pretty gray Haida, whom I could pet and handle but who was smaller, would lose.

Then one evening on a hunch I trudged up into the far woods. And here were two swirling forms leaping and playing, crushing down the knee-deep wild flowers. They were Haida and Yukon, having the happiest of romps. Nearby, Noluk and Shy, dark-eyed in the twilight, slanted their chins upward, looking at a squirrel in a fir tree. Alatna ran bowing to meet me, as I stood among flowers. It was a happy occasion. So—we would let Yukon and Haida work things out between them. Perhaps their aggressiveness was confined to the plaza, near us and the cabin; it was place-linked. The deadly power that lay always in my hands and might, perhaps, steel my eyes when the animals looked up at me—the power and maybe the necessity some day to kill—that was veiled.

An aggression that troubled me was that of Noluk toward

me. There had been a near thing with him and Shy once in the spring, when Noluk had seemed to feel this was it, the gratification he would have dreamed of, to bait me, perhaps worse. It was a curious fact that the heartfelt hate he had showed toward me since he was a pup was becoming attached to the side gate we used most and to the cabin door.

An incident in July connected bad feeling with the gate. While I was absent, acquaintances dropped in, a family with young sons. Returning, I found the husband and sons running and shouting above the pens, two things the animals had always feared and disliked—movement above them and children's shrill voices. The people said they had not entered the pens. The next day, after they had taken their private whirlwind and departed and I had driven off to the Lake, Syd, an old friend, arrived. Not an animal was in sight. As he stood in the yard bolting the side gate, Noluk came with silent speed and bit him in the bend of the knee. Syd showed us the bite when we came home. It was red, purple and livid white. We thought Noluk had been stoned, perhaps near the gate. From now on he hated not only me but the human race. And the gate was a releaser.

Two incidents fastened bad feeling on the cabin door as well. While I prepared dinner for old friends, their son, standing by the wolf hole in the door, playfully teased Noluk and Shy, outdoors. The wolf-dogs made short furious rushes toward the door. They would have killed the boy if they could have got to him. Before, in my distress, I could summon a kindly way to stop the angering pastime, the boy's mother stopped it. "I could see they weren't taking it like dogs," she said seriously. "It wasn't play to them."

Another time two fish-and-wildlife men called to learn whether one of our animals had escaped: a wolflike animal had been trapped down near the Springs. Noluk and Shy were in the yard, already brilliant and big with rage. "That animal wasn't anything like these," said one man, impressed and admiring. Both men began taunting the pair through the closed screen door. For the men this was a thrill like that of watching a thunderstorm at close range but from perfect safety.

This time I found my tongue promptly. "You're making these animals more dangerous," I said coldly. "I have to live with them." This was a new idea to the men. They were glad enough to creep across the table and out the escape window, not braving the storm they had stirred up.

It did not lessen Noluk's hostility toward me that I blocked an escape dig of his one day. There was no doubt of his purpose. For one thing he was radiant—happier than I had ever seen him. But what was chiefly revealing was the direction of his steady, happy gaze when in pauses of digging he looked out at the hillside toward which he was driving. The dig was actually under the fence and had only a foot upward to go when I managed to get him and Yukon, his pal and assistant, into the big pen and shut the gate. Then I overended a big rock into the point of the dig, snipped wire and laced the two ground aprons of fencing together. He had gone successfully between them at a stretch of about four feet where, as I found, they had not been laced.

Revealing, too, was Noluk's steady somber gaze out at the hillside as he lay by his thwarted dig on returning to it. He was cheerless.

In September a radical change for us occurred. The quartz at the mine ran under forty feet of "country rock" and Cris moved his machinery to a distant mine, returning home only on Saturday evenings. I faced my first fall and winter virtually alone with the animals. Now, if trouble arose, there would be no help for me until the weekend.

I still walked with the animals in the big pen—it gave pleasure to Alatna and to me. But I distrusted Noluk and kept close track of his whereabouts, thinking that I could protect myself. (I did not much fear Shy, though he sided with Noluk.) I still did not understand that Noluk's hostility was localized, was hostility-in-a-place. I felt that I should stock the Wolf Den cabin on the bench for a siege, in case Noluk prevented my return to the main cabin. I should carry in wood, lay in matches and emergency supplies.

In the midst of this tension a beautiful fact emerged: Alatna still counted me among her pack that she would defend. It took Noluk's hate to reveal this devotion.

One October evening, seeing Black Beauty nearly leaping over her fence, I shouted to scare her back. Instantly I saw that Noluk had taken note, but I was backed against the cabin logs and safe, I thought. But starting out the side gate soon afterward, thinking of no harm, I felt a touch, there was the clop of jaws at my leg. Noluk had rushed me the minute my back was turned.

This was nothing. What happened next—open to be seen, yet so quick it might as well have been veiled—was that Alatna flew at Noluk, not mad—when does a wolf get really mad?—but disapproving. She threw herself rebukingly crosswise over his back. Then the next second you would have said that nothing had happened. She was shouldering and sidling away sunnily with Mr. Shy in the departing throng.

She actually checked Noluk's aggression toward me one day. Her action was a trifle of the kind so remarkable one almost doubts oneself, yet it was unmistakable.

In the melting sunny snow after a blizzard, several animals were near me, on both sides of the gate. I was aware that Noluk stood still, looking at me with bright, hostile eyes. Alatna ran up, bumped his nose and ran on. But—wolves don't bump each other unless deliberately. Was Noluk more hostile than I supposed? Intending hostile action?

I feel sure now that he was. The brightness of his eyes indicated it. Usually he gave me the dull inward look of passive hatred. Probably Alatna had read the signs more clearly than I did: the bright eyes, the stilled tension of stance, the orientation of gaze toward me. She had acted not with the yowling, tearing, fangs-open ferocity of myth wolves, but lightly, deftly—and adequately. Her bump had sufficed to deflect Noluk from his intention.

But his hate remained. Trouble that the wolf matriarch could not avert was bound to arise. It grew depressing day after day to meet Noluk's lifted snarl, his unmoving hostility, yet always to give him pleasant talk.

In spite of trouble there were inlets of pure joy.

Before sunup one November morning, seeing the wolf-dogs all at the fence and looking toward my bedroom door when I rose, I threw on a bathrobe and went at once to feed. Two dripping pans of cornbread were ready to cut and give.

On the plaza, in the cold pure still air, there were the small activities of feeding. First Alatna packed off her piece of cornbread to bury. Then everybody disdained cornbread, but clustered around me to eat bites handed out by name. Noluk's jaws opened with a smack like saying Thus, when he stepped forward to his name and bite. There was the perfect fit between my skills and the animals' responses. I knew what the animals would do if I did this and for once it proved out.

The birds' turn came next. The frosty gray lace of the aspentops was marked by fourteen Steller's jays, playing "musical chairs" excitedly as I filled their feeding stands. "Here, bird," I called in my purest voice. They cawed and flashed down dark-blue, with decisive crests, to eat. Clark's nutcrackers came. Juncos hopped on the ground. A raven crossed low, looking for his food, the sheep carcasses in the big pen.

After all the feeding, when at last I was perfectly ready to eat my own breakfast, Alatna wanted in. She so rarely got in nowadays, with the outside dogs pre-empting the cabin, that I yielded. I put the dogs out and let Mrs. Troublemaker in. As I sat down to breakfast she stood on the table, her warm head at my forehead. She accepted a buttered hotcake from my plate —which she did not touch—and jumped down with it cheerfully. I sacrificed my bacon for multiple bites to all the animals, standing enchanted at the open door looking in or stepping in.

Then I lured Alatna outdoors with fragrant shampoo on the deck to roll in, and not to deprive dark Kobuk, silently but observantly hurrying along his fence, I took a hotcake and a perfumed rag to him and tossed them. (I no longer entered his pen to play, he had become too unpredictable.) The rag missed, landed bafflingly on the overhang wire. Shy, foamed up and snarling, rushed along the fence. I managed to snatch the rag and hurl it to Kobuk, who caught it and sailed off—he was foamed up too.

Then—then!—breakfast. Stillness flooded the glen like sunlight. The mountains stood up into the pale sky and stillness. The bare aspen trunks in the big pen below the picture window were streaked white with the first sunlight itself.

"It's worth it," I thought.

But by that night my constant question recurred, the question bearing on the success of my work here to keep Alatna and her companions at least partly happy in spite of captivity: "Are they sad? Does the weight of that feeling overbalance joy for them?"

Then—always the unexpected—Cris returned from the mine, now in mid-week. The wolves heard his truck first. They knew its sound as well as the dogs did and heard it sooner. It must have been still out in the meadow, clear around the bend beyond the beaver ponds, when they sensed the vibrations in the air and knew unerringly that this was no stranger coming—else they would have fled; this was Cris coming home. All noses toward the sky, on the lighted plaza, the great choiring resounded in the blackness of the mountain-walled draw. The dogs barked. I ran to meet Cris. He had come to bring deer trimmings and does' heads from hunters and their butchers. He left at once to return to the mine.

At dawn Alatna got one head. I gave it to her alone, nothing yet to the others. Like a halfback she doubled, outstripped all down to the spruces and there risked no more running but resorted to law, to wolf law: she laid down the head, and the young animals laid themselves around her and touched her head with their forepaws, asking in on the treat. She ate. They returned to me and now each got a piece of deer meat.

That evening, no let, they feasted on the total scraps. I gave Kobuk and Black Beauty their portion in a carton, for which Kobuk deserted the meat. He began to tear it up. Black Beauty's squealing play with him summoned all the rest, enviously watching this big thing of the torn box.

Black Beauty threw her front paws up and forward and the rock came to meet them. She sailed over it out of sight, came to view, sailing up and over and around rocks toward dark Kobuk, who stood, tail horizontal, panting and "smiling," awaiting her. He threw a foreleg over her neck like an arm. She pivoted on her forepaws and flung her hind part around so that she faced him. He tossed his head aside from her snatches. Once the little witch, in mad running that old Kobuk could not nor would try to keep up with, actually sprang to his back and rode him like a monkey for an instant.

Then Shy and Kobuk charged. Shy leaped at the fence, foamed up. The two big males charged back and forth, doubling like a flash, barking hoarsely, tails waving. Once Shy in desperation twirled completely around in mid-air.

If captives they did not know it now. They bathed in emotion.

I had not realized the shadow Noluk cast until I breathed with elation at his death; a weight was off my back.

What had brought on the death after my long anxious living since spring, when Noluk and Shy threatened me, was that Noluk attacked me. That was on a Friday. It was a queer experience in one way, in that its suddenness cleared my mind of some details. I knew I was coming up to the plaza from the big pen and that the animals were with me. But I did not know what–if anything overt–triggered the attack. I was nearing the door. Noluk came from the far side of the yard like a shot and bit my arm. I fell; both knees were bruised. The fall startled him back enough so that I got to my feet. The force of the blow had swung me around. I backed toward the door, saying quietly, "No, Noluk. Don't do that," and waving my arm toward him. I felt the air of unexpected death around me. I managed to tear off my jacket and wave it. He was crouching and springing, holding off a bit because of the jacket. I backed to the wrong side of the propped-open screen door, had to step forward and back into the right side. I did not know whether I would make it.

Indoors I stood for a minute not knowing what to do, not bewildered, just off-stride. There was crashing feeling, not expressible. I wanted to cry but did not. Through the door window I saw Alatna jumping on Noluk. She had not tackled him to save me, I thought, but she was showing him her disapproval. It was touching. Then, dreamlike, I remembered the form of a wolf between me and Noluk as he charged head on, eyes on mine, to bite—this high brownish back moving as if casually between me and him.

I did not go out the door again that day, though I opened it and fed from my hand as usual. Noluk took bites of food; I felt his warm tongue, harmless. His eyes were clouded—sharp with suspicion but dull as to light.

On Saturday evening Cris came home and on Sunday morning put Noluk to death. It was a sad, mixed day. Cris made a perfect shot, instant death. But it was terrible because Shy and Haida stood within feet of Noluk. They fled. Afterward came the great doomed cries from the woods. "Which one?" inquired Cris in a low awed voice. I went to see. It was Haida. I would not have known her voice. It was the voice of total woe and desolation. "Was he her 'husband'?" Cris asked. I thought not.

What had been the last straw for Noluk? I did not know. When I went outdoors before the attack, he had never been more friendly. He smelled my hand, gave it a casual lick and stepped off wagging his tail. I went cheerfully with the animals down into the big pen, feeling very hopeful that now at last I might succeed in winning him.

What should I have done? Would anything have made any difference? Probably not. He had lived in a queer, uneasy equilibrium, showing how potent and how opposed dog nature and wolf nature are. They seem alike. They are worlds apart.

Perhaps the subservience of kissing my hand had revolted him, and my nearness to the door had triggered the attack.

On Monday still the animals cried. Not the special grieving cries of the day before but generalized mourning and sadness.

There were strange new tensions and unbalances in the big pen. For one thing, Yukon, who had persecuted Haida for months—a reversal of their earlier roles—making her cower but making her life intense, abandoned her. Haida seemed lost. Yukon tried to do something about her changed position. I thought her season had just begun and that Noluk had been her intended mate, her fiancé, as it were.

She gave her attention to Black Beauty, in the home pen. First she fairly shrieked with bafflement at not being able to get at Black Beauty. Then she stood up backward in Shy's "arms," both animals looking at the two in the home pen. Next she did an odd, impressive thing. She lay motionless at the fence, looking steadfastly at Black Beauty, lying motionless just inside. That was all. But it was one of the deadliest things I had ever witnessed. I was about to bring a camera when I realized that to the eye the scene was static.

Yukon followed up by courting Black Beauty's male friend, Kobuk. (She did not court her mother's male friend, Shy, nor did he court her.) She laid her chest down on the paw-beaten snow outside Kobuk's fence and pushed herself along luxuriously. She rolled. He went wild, leaping high against the fence to get to her.

I was relieved when, a few days later, she once more put Haida to flight. That afternoon the animals seemed peaceful and even sunny. They trooped to the door. Alatna fluffed into the cabin, Haida after her. Mr. Shy stood genially at the door and yawned, a pleasant sign of relaxation.

Yukon, too, came indoors and looked around with her mysteriously hard sharp glittering eyes. She was as beautiful as an enchantress. Her mingled dark fur was full, her tail a flowing plume. Her clear eyes were set in black slits in her dark steady face. An instant quailing rippled over her hair-trigger nerves at any act of mine the least out of the ordinary. Yet her warm tongue was quick to caress my hand, which she would not allow to touch her. (Pretty gray Haida, a smaller animal, had long since elected to let me pet and handle her.)

The weekend following this peaceful afternoon was tragic. When an acquaintance camped at the foot of our draw to visit us and observe Alatna, Cris and I talked over my lonely and perhaps hazardous situation: Kobuk had become dangerous and unpredictable; I feared him. I feared going into his pen to feed and had considered other ways, none practical for long, of feeding and watering. We agreed to ask the visitor to put to death the two beautiful, ruined animals in the home pen, Kobuk and Black Beauty. This was done. By accident gentle Haida also was killed.

That deadly night, when the animals were still in terror from the gunshots, circling fearfully into my lantern light on the bench and away into the dark woods, steadfast Alatna came up in the icy moonlight on snow and kissed me.

It was twenty-four hours before—daring, after countless hesitations—the survivors entered the home pen.

Yukon was as if wiped out. Her eyes were lightless. She moved listlessly, quailed easily. She was killed in spirit. She had lived for interaction. It had been taken away from her. Her brilliant-eyed dominance of pretty little Haida, her brilliant courting of inaccessible Kobuk—all was taken away. Her heart was as if dead.

At night the animals mourn-howled endlessly. They were lonely. Their numbers were too few. We had fought overcrowding. Now the balance had swung too far the other way.

The days came on time, smoothly, without friction. The sunlight colored the granite sheers against the blue. It came down the snow among the pines. Birds busied themselves.

In January, at dawn of the day after the worst storm in a decade, looking upward from our shadowed draw I saw the first sunlight like yellow glass on the east-facing rock. I could have cried with defeat because there was joy again, in spite of death and harm. As if nothing mattered but that we go on.

Later that morning Alatna came up to the plaza in the blazing ice sunlight, imperious and impetuous as of old. She had weathered the emotional storm better than Shy and Yukon. I realized that I trusted her as much as a sister. She might get peeved about something but she was perfectly dependable. There was no hidden pool of distrust of me, or of meanness in her, to be unexpectedly tapped. I was on my guard a bit with Shy, more so with Yukon. I knew Alatna. She was sweet of nature clear through.

In turn, she trusted me more deeply than ever. The old days when we had doubted each other were gone.

That evening I closed up all duties in the cabin early and with a book retired to the bedroom. A howl sounded. Duty and weariness conflicted. The former won. I threw on a bathrobe and trudged down into the twilight snow of the big pen to hunt and console the wolf.

Crunch, crunch behind me. It was Alatna, rushing over the snow toward me. She was as beautiful with delight as ever, as if life and death and tragedy had not passed over her. Then she did this. Right in front of me, within arm's reach, her back toward me, she took a tall leap, perpendicular, and at its peak tossed her head around and glanced over her shoulder at me.

I was glad I had come down. Rarely had she done this—made this special, lovely gesture. Wolves do it for each other. Also the wolf-dog Noluk had stood up this way in Shy's "arms" and looked with bright bewitching eyes over his shoulder at the elder wolf-dog. Eyes such as he never showed toward me.

The next morning there came a signal from the future—a sign of new beginnings and of changing status for Alatna.

I cut a chicken into three pieces and took them to Alatna, Shy and Yukon, who were in the yard. But when I saw that the first two pieces given were not much regarded I was about to take the third piece to the outside dogs. At this instant there was a yelp. Yukon had caught a baby dog pup's paw through the fence and was hanging on. I threw the third piece to distract her. Instantly Shy charged me, foamed up. Alatna threw herself between us. She bit lightly on Shy and got him down. Black-eyed, she punished him by keeping him down, submissive, for a while. She had never feared Shy. She *had* feared Noluk.

But for once it was not Alatna's protection of me that mattered most. The point of importance was Shy's defense of Yukon—his mate.

20 *Father Wolf-Dog*

YUKON did not use her mother's good idle den but one all her own, the dark space under the rear of the Wolf Den cabin on the bench. For a few days before giving birth she was gentle, patient, "smiling," with gentle tail-wags. She was not under stress as Alatna had been with her first puppies. Alatna had been alone. Yukon had the actual physical help of Shy and Alatna.

As she lay in the den with her newborn puppies, Alatna, over on the plaza, took from me a favorite food, broiled beef liver. She hesitated as if uncertain what she was going to do. Then she became certain and ran off, gay-tailed, toward the trail up to the bench. Halfway up the trail Yukon came fawning and crouching from her den to meet her, and Alatna let her have the liver. It looked as if that had been Alatna's "purpose," to take the meat to Yukon. A wolf is naturally a giver.

I could not feed Yukon in the den as I had fed Alatna; she was too wild. I knelt near, holding out the food, and she came from the den to take it. The first time I offered her food without Shy penned away, he came up and gave me a sharp warning nudge in the rear. He and Yukon were a team.

Taking meat to her once, I passed him lying in the shelter.

Cris told me later that as soon as I had gone by Shy went and brought Yukon, who was down the avenue, and that she followed me up the trail. Cris would have warned me if he could. Near the den I stopped, peering toward its blackness; I would not go closer if Yukon was not within. Suddenly I felt the close clutch and the pain—a bite. I turned to meet Yukon's surly eyes. She had retired downhill a few feet. I knew it would be safer to go but I risked it and stood still, speaking gently and holding out the meat. Eyeing me distrustfully, Yukon stepped up and took it. After my honorable conduct toward her puppies on this occasion, Yukon trusted me more.

She had six puppies, all males but one. Two were stolen and killed for population control.

Shy, Yukon and the four puppies were building something not seen before at Crag cabin—a wolflike family, complete at last with "Father Wolf." The family would bring a new role to each adult, including Alatna, but it did not occur to me that she would ever change. I did not realize that she would have to decline in rank, become only the dowager queen. But for now, and all through the time when she would normally have been pregnant, she retained her old supremacy and confidence.

The game warden brought us a dead fawn. I dragged it only as far as the terrace, where Alatna took charge of it. She was wonderfully set up to have a wild carcass to guard and give; she did not eat.

First she savored the guarding. Yukon came in vain for meat and vanished. Next came Shy. Confronting him, all at once Alatna's face took on a dainty, fanciful look, as if she had dressed up, put on a necklace. Her bared teeth looked like an ornament. She stood absolutely motionless, holding her teeth bared. Shy was a statue too. He stood cowed aside, head turned away low. The pair held the tableau for a minute, then Shy gave up and left. Alatna did not seek shade but lay all afternoon on the sunny terrace with her prize.

Yukon returned at sundown. She and Alatna and the fawn were in line. Yukon came toward Alatna in a perfect heat of radiant submission. You would have thought she was touched and overjoyed at coming to this lovely protectress and honored "personage." But she never for an instant relaxed single-pointedness toward her goal but drove on low toward the fawn. This time Alatna, like King Ahasuerus with Esther, tolerated her presence.

Yukon tore at the skin. She tossed mouthsful of gray fur into the air. She glowered at me when I stepped outdoors with the camera, and she dragged the carcass down over the wall and clear on away through the gate into the spruce shelter.

In March a lost puppy caused me an odd, pleasing little adventure of my own. I was wakened in the dead of night by a pup crying Lost. Its mother, Yukon, was the one that easily became hysterical, and when she was, she knew how to escape. It was urgent to get the wild untouchable pup back into the pen—how I did not know. I did not even know where in the steep forest in the darkness the pup was. The cries seemed to come from the mountainside.

I snatched on boots and jeans and with an electric lantern went out to hunt. The wails stopped. All the animals, including the lost pup, must have been standing silent, fascinated by the signs of my progress—the waver of yellow light on the undersides of black boughs and the rustle of my boots in the snow. The air was icy-still and pure.

Half bewildered yet by sleep, I crawled by the beam of the lantern over great frozen waterfalls, mounding away into the dark forest ahead and sliding downward at my side, fifteen feet of frozen white. During the winter the cascades had frozen and swelled in mounds out into the deep forest. I did not know where I was or where to cross. It was mysterious and delightful. Except for the lost pup I should not have been out here in the night forest among these shadowy ice mounds.

I did not find the puppy. Silence prevailed until I got back to the bedroom, then the wails broke out again. I located their source now. The puppy was safely inside the pen after all, down by the brook at the foot of the mountainside. This was the confused night when the puppies first moved down off the bench into the home pen and, halfway between, this puppy had lost its nerve.

In May, when the puppies were nearly three months old, Alatna was still the confident matriarch. She joined happily as usual in the wake-up howl one morning.

Shy, on awakening in the dawn twilight, remarked "Ow." Silence. Presently another good-humored, tendentious "Ow" from Shy, a brief hollow sound like the voice of shadow itself, indistinguishable from earth and water. I went down to the shelter to say good morning. They were all there—Shy standing up and stretching, holding his position so that I could press down on his back again and again, just slightly intensifying the stretch of his fine muscles; Alatna in her hay bed leaning up luxuriously to kiss me; the soft lamb's-wool puppies off at one side with their dark slim mother.

They all assembled outside the shelter. Then came the full delightful howl, all amiable. Alatna put her throat up, half closed her eyes and sang. The pups considered what everybody was doing, then irresistibly, at seeing Mr. Shy's big muzzle go up and his mouth widen, they too, sitting or standing, put up their muzzles and piped their coloratura yowls. All moved around among themselves. Every tail waved, all ears were tented. The pups' ears went up, then down as they looked toward the adults and howled again. Kisses were given inside this mouth or that—pup to pup, pup to open-mouthed adult, a quick caress of the tongue from an adult to the face of a pup. Every eye beamed with contentedness and lifted spirits. One pup in the movement found himself facing away from the group and flattened his ears sweetly back of wide-open mouth as he continued to join in and howl.

Then it was over and Shy terminated it wolfishly by uttering a few short hoarse scolding sounds. He was still thronged by the excited pups. They did not move away from him or from his jaws but accepted mouthing happily. All was genial. I hurried up to the cabin, calling as I went, to confirm the general good mood with cornbread and cream poured into milk.

The sun topped the mountain, the deep glen was still in shadow, but up here on the hard plaza the small "furs" were backlighted, lounging over on the edge against the black shadow. The big wolf-dogs came into the sunlight beside me. Alatna stretched out, warm-furred, on the plank table back of me, against the log cabin. I tossed bites of cornbread, and half a dozen Clark's nutcrackers gathered around, moved to speech by the stresses of their own situation as yellow chunks fell invitingly and the birds were thwarted by the young big-legged furs. One bird skittered down the side of the plaza after a rolling chunk. So I called impartially, "Here, bird. Here pup," and spread the tosses.

Could I have shared any of this pleasing dawn idling? It would not have occurred if anyone else had been present: I could not share the animals with anyone.

It was soon after this morning that Alatna changed. She had a pseudocyesis again, not a full one, perhaps because of the presence of the puppies. But still the old tide and need and longing were in her blood. Afterward she was passive, she lay around. Her crippling seemed to hurt her more, impede her more than in other years. She did not "fly," excelling all others in speed. She limped after the rest, seemed to prefer not to walk or run. My dear Alatna.

By the time the puppies were four months old she had given up expecting to be number one animal. Formerly when I went outdoors she had come up demandingly, imperiously, sure of her precedence. Now–I hated to see it–she stood down by the home-pen gate, looking up toward us as the others gathered around me on the plaza. Or she hobbled off toward the big pen, going off alone.

As for Yukon, she was quiet and secure. She had fun with her puppies. She teased them and played dancily with them.

Like Alatna, Shy changed but in the opposite direction. She was moving toward the margin, he toward the center. He was Father Wolf. Again and again in the course of the day we saw him hurrying off in a businesslike way, "proudly"—head high, tail high—carrying something toward his mate and young, such worthless thing, bone or dug-up meat, as his captivity afforded him. It was the mere shadow of the important work of a father wolf; it made me sad. Always I hated captivity for these animals.

Shy ranged off alone a good deal. He was, in point of fact, different from the others. He was the lone adult male, leader, father. He seemed to have a need to range, to leave the young and go off as if proudly and independently, as if in this pen he still had business-of-life. He asserted in these two ways—by aloneness, by contributing—his authority.

Also he had a daily need for asserting dominance over the most important male dog, namely, Toklat, leader of the outside dogs. As Shy faced him, his stance of pride was like that of a splendid stallion, head reared, eyes bright, tail high. The fur ruffled up a bit on his shoulders but also perhaps his tail waved slowly in pleasure. He treated Toklat differently from Kobuk. With Kobuk, his equal and rival, he had conducted the rushing power charges. Toward Toklat he showed disdain. He scratched backward with his hind feet, stepped aside to the rock wall at the side of the plaza and urinated very deliberately.

The dogs ran away once and straggled home days later, one or two at a time, ribby and limping. Probably they had followed a deer and got lost and had just rambled around until they found home. It was five days before Toklat came home, alone, beat out. Shy's behavior interested me. He had seemed to miss his prestige bouts with his big opposite number: he had watched toward the road and garden for him. When at last Toklat did appear, Shy and the other wolf-dogs were all excited. They wowed, waved tails, gathered along the fence, stayed watching even when tired Toklat went into the dogs' hay shed out of view. Shy performed his superiority routine, scratching backward and urinating. He was exhilarated all the rest of the afternoon, tail and head high.

The change and decline in Alatna were not a matter of her exclusion by the family nor of pique on her part. The family still cherished and looked up to her, and she cared for them.

Shy and the puppies and Yukon waited at the open door one morning for her to emerge from the cabin. She had lain on the bunk regaling herself with sprinkled perfume while I ate breakfast. Shy's ears went down for a moment, his eyes beamed when he glimpsed her coming. This or that pup licked out its tongue as if in readiness to kiss her. She sallied outdoors, accepted the thronging with regal pleasure. Tails wagged. Shy picked up his paws and leaped right over her from a standing start, like a bighorn lamb over a ewe. All followed her down over the terrace wall and off into the woods.

Alatna for her part cared a great deal about the puppies and threatened me over one of them one evening. I took supper to her in a hay bin in the vestibule of the home pen as, after thunder, rain and hail, she still lay there snug. (The puppies were dark with wet from scuffling in the rain.) As a puppy reached into her pan and I took the chance to pet him while he was reassuringly close to Alatna, he shrank and looked up at me. And Alatna looked half up and growled. It was menacing for her not to look clear up into my eyes. Her eyes were cold. I was not to drive the pup from her. (But afterward, after all additional tidbits had been distributed and all pans licked, she smiled upward with her eyes and jumped up and kissed me. Was she making it right for that unfriendly threat?)

I recalled a recent occasion when she might have been expected to threaten but had taken a passive tack. It was on a morning when the wolf-dogs had dug out of the yard pen into the dog courtyard. All were appearing inside the pen and out of it, so I recklessly threw open the gate to herd them back in, forgetting that Alatna alone was still in the pen. She came out and I started toward her, arguing and rebuking. I meant to force her back into the pen. And she crouched, sitting wide on her haunches and crying, her ears flattened, her head twisting. She knew what I was going to do—and I did it and she let me do it—but she was protesting and crying about it. She let me paw-drag her, still sitting, back into the pen. She was not hostile but submissive, though vocalizing.

But when driving a puppy away from her was involved, things were different. Dethroned, she was still matriarch at heart.

On the last morning of July I reviewed her situation during the past two months. I did not suppose it would change.

"She is old and hopeless now," I thought. "She needs a fine big old male dog or wolf to be her equal and give her status. Then she would have a 'purpose' in life again and zest." The difficulties of providing such an animal would be hard to work out. He could not be introduced into the pen with Shy. Shy would fight him and, being young and in his prime, possibly wound or kill him. If Shy were removed, Yukon would be without a mate.

Shy, Yukon and the puppies were a whole: they composed the family. Their adequacy and trust and interrelation were apparent. They lay together, took an interest together in the mouse that seemed to live in the rock wall at the side of the plaza. An invisible yet perceptible sufficiency among themselves surrounded them. Shy dominated the pups, asserting this dominance by a growl or a snatch at an intrusive muzzle. The puppies cooperated unconsciously in the invisible union, that of lord and lesser. They were not subservient, just compliant and engaged. The six wolf-dogs formed a vital unit, this family.

And Alatna? She did not intrude. She stayed much by herself, over on the bench in the dry dusty root hollows under the big fir. Her physical strength seemed to decline. In the old days she had played madly to the gaze of her own puppies, who were unequal to her in speed or joy either.

Her old tricks to "stay on top"—not used consciously but as the natural flowering of her impetuous zest—did not suffice now. She was a supernumerary. She never, never tried to dominate Shy any more: he was lord. It had been a seesaw relation between them for a while there, as he had felt his surge of new power—Yukon, the pups, his authority with them, their reciprocation to it, independent as they all were.

Alatna was respected. She was alone, not needed. Her eyes were not happy—terribly intelligent but seldom sunny as in the old days. On this morning of my meditation she had come into the cabin, once the source of innumerable possibilities, had looked around and soon gone to the door to be let out. I praised her and she went out with her skimpy tail elevated at the rump in the old happy way. But she had barely nosed the perfumed rag I offered though I had soaked it with her old favorite, Chantilly.

The perpetually coiled spring of will-to-escape—to big liberty, to hope, to wolf life, that lay on the mountainsides visible all around through the wide-meshed fence—that spring was slack.

Yet it was not all quite that simple. Alatna had the respect and fondness of the family. Yukon had recently apologized respectfully to her with a kiss for an unintentional nip. Alatna did play with the pups. She came up over the terrace with sunny slant ears to kiss me when I came home from the Lake or Springs. Perhaps I had changed toward her. Still I felt, "She is licked, barring an infusion of money to make a new pen and an interchange pen and bring in a potential mate for her alone, yet allow her to re-enter and range the old big pen, where her pals are."

Alatna! Tundra girl. This is what I often remembered, an incident that happened on the tundra one day when she was young. Gay graceful quick Alatna, full of games, usually led the pack. On this day she and the other wolves had played ring-around-the-rosy around a low willow clump on the river bar. Finally one or two took to cutting across the top and coming in, teeth foremost, at the very rump of Alatna, who was leading as usual.

Cris and I started on downriver. Alatna, sure of our direction, led off, followed by the others. All but the last two wolves were out of sight when Cris abruptly changed direction and started toward home. We called back those last two. They followed us but looked in the direction Alatna had taken and cried worriedly as we went north while she had led out of sight west.

At last she led her two truants back to us. She came racing, ears back, paws flopping gaily to the sides. Cris knelt and she flew up to him and kissed him.

That was all. A runaway and then return to us on the tundra. But it was Alatna for me. All that had happened since, of death and joy, only amplified it.

21 *A New Order Emerging*

On the very morning of my lament about Alatna her eclipse ended. She took charge of her diminished life and began action.

The morning's diversions had been over for half an hour—running to welcome me as I entered the yard, receiving breakfast, a broom, a perfumed rag, coming into the cabin, etc. The puppies were on the plaza, Alatna was over on the bench alone. From there she uttered a sound new to me, a short descending hoot, "Oo-oo." It was a summons but not the "puppy call." It was a wolf yoo-hoo. Evidently nothing happened, for Alatna began to elaborate it, not into a howl but into short variations on the call.

I was starting to go to her when something did happen. The soprano chorus of puppy answers broke out. As I reached the screen door I saw only the rumps and flying plumes as all four puppies slanted down over the side of the plaza to forgather at the source of the call.

In the old days Alatna would not have yoo-hooed from a distance: she would have been in the midst of the puppies and they would have been her own. Still the call marked action after her long passivity. Perhaps the near close of Yukon's nursing cycle heartened her. Perhaps it had taken this long for her spirits to revive after the sterile defeat of pseudopregnancy.

That evening she performed another new action: she courted Toklat, a potential exogamous mate. (This early in the game the courting was in play.) She came indoors when all the dogs were lying in the cabin. Mr. Shy stood outside the screen door and watched black-eyed. What was striking was that both wolf animals talked incessantly; the dogs, except Toklat, were silent. For her part Alatna was crooning to all the dogs. Mr. Shy was saying Mow-wow-wow—not in the least a bark but rather a modified howl.

Toklat bit Alatna's muzzle lightly, growling at her and barking at Shy, who reared up, paw on screen. Toklat's biting intensified the crying and talking by Alatna. She was submissive. After all, her only mate had been a dog and he had done just this—growled and bitten her, and out in the pen with her own family no one dared bite at her or dominate her in this way. (I had to put a stop to the affair: Shy was becoming too dangerously black-eyed; he could have come through the screen door.)

A few days later Alatna made the tremendous emotional adjustment of acting submissive toward Shy, whom she had always domineered over until recent weeks. He stood on a travel box, a status symbol—actual elevation conveying prestige elevation—and had a lordly bout with Alatna, who was on the ground. She enjoyed it. Her eyes smiled as she bowed, entreated, reared her head, persistently offering to kiss his muzzle. It was an emotional workout for her. Probably she craved it. How often in the old days she had experienced being female and entreating toward a lordly male.

That evening she performed a new surprising little act toward me. I was walking on the hill flat, the bench, when she came up from behind and standing up beside me laid her good "arm" over my shoulders and kissed my cheek. For the first minute the gesture felt so simple and natural that I thought nothing of it: a friend had overtaken me and laid an arm across my shoulders. The next minute I felt awed and hilarious: a wild-born wolf had done this thing. (A wolf often throws an arm over another's shoulders, but Alatna had been obliged to stand on her hind feet to do so to me.) She was gay because I had come over here at twilight, play and gay time. She had run and twirled her head, bowing, with more spirit at this late date than all of Yukon's pups put together.

In September, seven years after giving the "pact" for Alatna, seven years of my life, I almost killed her by accident. (I did not think about the pact now; a door was closed between me and it.)

On the walk one afternoon the dogs ran away again. An acquaintance from Iowa was visiting us. Returning home toward evening I told her why I was so disconsolate. Carelessly I added that the only help I could give the dogs would be to drive up to an open ridge facing the mountainside—toward which they had run—and honk the horn to orient them (as if they would be disoriented this soon after leaving). That was out of the question: it would terrify the wolf animals.

The woman from Iowa gave me a look of condemnation. "Well!" she reprimanded. "You'd better go honk your horn."

As if my love and work and certain knowledge were of no consequence, I complied. In my new yellow four-wheel-drive Scout I drove down to the aspen flat at the foot of the draw and in roaring compound low up to the open ridge. Turning off the motor I sat facing the darkening, forested, impassive mountainside and honked the horn again and again, aware that each prolonged piercing blast, echoing strangely from the crags, would be increasing the terror of the wolf animals.

At home again I strode past the lighted cabin, indifferent to dinner preparations within that I should have been forwarding, and hurried down to the avenue, calling. Not an animal was in sight. It was dark twilight. There was a crackling on the hill-side above me. Alatna burst from the brush and ran toward me. As she neared, all in one swift movement of running and sitting she threw herself on her haunches and skidded up to me sitting, her head and ears lowered. She was crying passionately. Her intense reaction revealed the depth of her fear over the unprecedented noise.

After dinner there came mournful crying from the woods. Going to investigate, I found the animals still in a state of apprehension. Too many things were wrong—the dogs gone, the horn-honking, and, diffused over the woodland, the unfamiliar smell of an oil heater started up for the first time in months. I shined Shy in the garden and went through the urgent routine of getting him back into the pen. Yukon was going in and out of the pen by mysterious means of her own. I shined her out on the ridge. When she was in the pen I tranquilized the animals with sleeping pills, overdosing Alatna by accident. Probably Shy had discarded his baited meat and she had got it.

Fearing for her the next morning without knowing why, I went in the first gray light to find her. I found her only through the direction taken by the other animals. She was in a bad place, a tight triangle formed by two great rocks and the fence. She lay heaped against the fence as if she had been struggling to escape when unconsciousness overtook her. When I started to move her heavy, resistant body, Shy evidently thought I was not doing the right thing by her. He was prepared to bite me. I talked placatingly to him and he accepted my actions but followed me and watched sharply, along with the other wolf-dogs.

I tried to rouse Alatna. I lifted, dragged and half-walked her up to Coonie's old tree. I supported her until she stood shakily. The pleasing thing was the delight of these social animals at seeing her stand, finally, on her own feet. They all ran up, tails and eyes happy, to kiss her face. The half-conscious wolf flattened her ears in instinctive friendliness.

I laid her down on the dry duff. The wolf-dogs went back to their own sleeping place. Alatna was trembling with cold. The sky had paled but it would be cold down here in the forest for another two hours. I lay down curved around her spine and stretched an arm and half of my jacket over her chest. So we lay in the stillness.

Her trembling stopped. I recalled seeing denim jeans and shirt, stolen by the wolf-dogs, and went to find them. One of the young wolf-dogs woke up and followed me back to Alatna. I spread the clothing over her, gripping it and scolding softly until the wolf-dog let go of it and went away. I left Alatna lying covered and went to prepare breakfast, thankfully sure that she would come out all right.

In October two major changes occurred for Shy, the first of them good. Musa, biggest and chief of the young wolf-dogs, became his courtier. Submissively Musa begged attention from Shy with flattened ears and bright eyes as a proper wolf should, and Shy stood still and lifted his lip and growled, as Kotzebue had done toward him when he himself had been Kotzebue's courtier. Musa had worried me because, though Shy avoided an issue, Musa had been rude and overbearing, had snatched at Shy and taken the elevated place, the travel box. Now there was this happy, ceremonialized relationship between them. Shy's authority was matched by Musa's submissiveness; both animals were equally happy and free and assured. Their conduct seemed a gay rite, highly enjoyed by both parties.

The other change for Shy was grievous. Yukon went mysteriously wild with glee one day. Now I learned how she had been able to go in and out of the pen at will when excited. She climbed the fence like a squirrel, paw after paw, gripping and up and over. From either direction. I brought boards and wire and blocked her route, only to have her disclose another. She walked up the slanting two-by-four brace on the side fence and crossed the overhang wire on the main fence without much trouble. Each time I blocked one exit she found another.

I feared, as with Kotzebue, that at any moment the other animals would learn from her how to escape. To save them I put her to death. I gave her sleeping tablets in lethal number. She was happy almost until she toppled over and did not rise.

She wagged her tail, toured around incessantly, kissed her grown puppies. They kissed her. She kissed Alatna muzzle to muzzle. "Saying goodbye," said Cris sadly when he heard about it.

Shy was clearly lonely after her death. He would not join the young wolf-dogs. He went often to the farthest woods and mourn-howled. He could not be helped in his loneliness. He had had a mate. Their closeness and what they had still meant to each other, even now when the pups were eight months old, could be divined from his grieving.

22 *The Widowed Two*

I DROVE homeward alone in falling snow one November twilight. The firs along the Tarryall road were colorless and dark, the mountains ahead gray with snow. For twenty miles I had passed few dwellings and no cars. Turning off the earthen public road, I stopped at the first gate, got out and opened it, drove through and closed it. Three more gates, then I passed without stopping between the cabins of Enid and Victor. For two more miles, through other gates, I drove over the rolling pastures, whitening under tan grass, toward the dark mountain wall as if expecting to tunnel under it. Rounding the prow of our home ridge past the beaver ponds into the aspen flat, I shifted the engine of the Scout into low gear and drove steeply upward along the side of the draw, honking the horn—brief approaching sounds that were exciting and familiar to the animals.

Along the garden fence dogs' heads were thrown back. Over in the yard pen shadowy gray forms were coming up onto the plaza. Tails over there began to swing. Heads went back. When I stopped the engine I heard the uproar. I went at once to greet the animals.

I made impeded yardage through the dog courtyard, distributing mismatched greetings. "Doonie!" and my hand alighted on Elf's head. "Old Toklat," and my hand touched Buster. At the side gate by the cabin I made the interchange, safely unaccompanied, from dog world to wolf world. Alatna was just coming up onto the plaza. She paused beside the old wolf travel boxes—two of those in which, eight years before, she and her fellows had come from Alaska—and holding her neck straight out she uttered a long level "Oooo." It was not a common sound but one I knew: it was a special "to you" greeting, showing pleasure on the part of the wolf. Then she came on up to kiss me. I had entered the domain of one of the world's most social animals.

What happened next was not on the face of it very social. The dogs rushed into the cabin by way of the hatch in the bunk nook. I was closing the hatch when Alatna came indoors. This would have been all right except that Bakno, the only female wolf-dog, slipped in with her. (Bakno could not open the door by herself.) The room was in an uproar. Doonie, aided by Buster, attacked Bakno. Bakno cowered and skidded, no toe grip, on the linoleum. In the melee, trying to slap Doonie, I whacked Alatna. She did not lightly forgive that thump. When I got her and willing Bakno outdoors and apologized, she both kissed and nibbled my face and at the same time growled. Her eyes were black. She seemed poised on a knife edge, ready to go either way. As a substitute for punishing me she jumped on Shy.

He reacted in two ways at once: submissively he cowered, neck to ground, but his rump stood erect and he snarled upward at Alatna. He was not the underling she had formerly domineered over. He had known authority.

The major problem these days was that Alatna had changed again. After the death of Yukon she had reverted to being full matriarch, not dowager. She no longer yoo-hooed to the young wolf-dogs from a distance; she frisked and played among them. So much was good. But also she had given up her brief submissiveness toward Shy. She tried to dominate him as formerly.

This would be fatal to mating between them. Yet the widowed wolf and wolf-dog needed mates and their only hope was to accept each other.

"Those two will never–*can* never mate," I thought hopelessly, watching her jump on Shy. "They are doomed each to sterility from now forth unless I can drastically change the situation for both in the pen."

Later that evening there came such a sad howl from the darkness that I took the flashlight and went down to the shelter. I knew what kind of welcome I would get: the firm generous affection of the wolf would be a solid comfort. Alatna leaned up from her hay nest and kissed me. I realized that it was all right between us; she had been peeved for a minute over the slap, that was all.

I was aware of Shy standing beside me, waiting silently to be "loved" too. So I petted him. The other wolf-dogs, their eyes black in frost-whitened faces, lay in their nests so close I could have petted them, but they lay still only because they were sure I would not touch them. As to Shy, this animal had once had in mind to kill me, and I liked his shy proud wish for affection.

The next day he took special pains to speak directly to me.

Usually on the dog walks I kept a lookout ahead for bighorns and steered the walk away from them. But as they had been absent for some time now I had grown slack. On this day I saw them ahead on a rocky knoll just as the dogs saw them and gave chase. The bighorns fled to a high ledge where the dogs could not reach them. I called all the dogs to me except Buster, who would not come. I led homeward without him.

As I was preparing dinner I heard Shy talking to me. He stood on the deck table, alternately turning his head to look in the door window at me and straightening his neck in order to "talk," wowing his fanged lower jaw fast up and down. He was notifying me of an event of importance. I knew exactly what it was: Buster had come home. Shy had left his view of the event and come expressly to communicate with me about it.

I ran outdoors. Outside the gate to the roadhead sat the yellow form of Buster, pleased but subdued. I wished I could let him into the wolf pen, somehow to satisfy the wolf-dogs' excitement over his return. That of course was not possible. But perhaps my fuss over him made them feel the event had been brought to a satisfactory conclusion. I gave them a howl.

I cared for Shy almost as much as for Alatna. Or rather in a different way. He had been obliged to overcome so much in himself in order to tolerate and like me; I had to allow for the social distance-between-us of such elegant structures of ritual, of what he deemed his due. He wanted my attention, but on his own terms. These were that his pride and autonomy be absolute. It was a touching ambivalence.

Except toward Alatna, Shy held to his ways as chieftain. He would dance up to the fence with bright longing eyes to challenge Toklat, holding himself so proud and high that the soles of his feet seemed barely to touch the earth. Though I knew it was impossible for him to be up on his toes, I had to glance more than once to be sure. The chrysanthemum of fur foamed upon his shoulders.

He still had Musa as courtier. I gave Shy a beef tail one morning. Bounding down off the terrace, he trotted to the shelter, where Musa delightedly crouched before him, giving little tosses of his head, impulsive quick advances of a forepaw, all the while wagging his tail furiously. But the demeanor of his head, centered in his electric, happy eyes, so far outshone all else that one hardly noticed the wagging tail.

Shy meanwhile, doubtless with equal pleasure, maintained his own role of dominance–slow, high bearing, no frisking or effusiveness of the ears but, irresistibly, a slow pleased wag of his elevated tail. I could not think of any human analogy for this gay rite. It was conducted with so much delight and zest by both Musa and Shy and was so purely ceremonial. That beef tail was worthless; neither animal gave a fig for it. What was important was all this ceremony bubbling up and effervescing between them.

There was no compulsion that I could see on Musa. Even now, Shy avoided pushing him too hard. He avoided a clash, preserved his special, elegant role by withdrawing. There was a compulsion on Shy: he had to carry out his alpha role. And the way he and Musa behaved was almost exactly the way Kotzebue and he had behaved in the old happy days when he himself had been Kotzebue's courtier. With this difference, that behind Kotzebue's superiority there had been the solid thrust of real and physical authority: he had been ready for the issue if necessary.

I realized one day that almost everything I did had an unconscious reference to the animals. If I was quiet, it was not to rouse them. If I was noisy, it was to make them gay or excited. If, driving homeward from the Springs, I saw dark-blue clouds ahead, I wondered, Are the animals having a bad time under them? Nearing home, I felt dread at the thought of seeing an animal in the wrong place—Alatna or the wolf-dogs in the pastures. (People miscomprehended about things like that. Invariably they thought the dread was of being attacked. But it sprang from awareness of what would happen to the animals if they got free. Always there was the memory of Tundra and the other wolves, moving eastward into death, civilization.)

Listening to the dogs bark at night, I wondered, Are they barking at a rabbit or is a wolf in the garden? Many times I had peered in the moonlight down toward the garden to see if a wolfish form moved there.

This was my second winter mostly alone with the animals. Their companionship was rich, but I missed people. Because of the bad roads visitors did not come during the winter. Marie, lacking a four-wheel-drive vehicle, could not come in stormy weather. The three miles of road over the cattle pastures were not maintained. I confessed my loneliness to Enid once. Her blue eyes filled with tears. "If you get lonesome, honey," she urged earnestly, "you come on over and we'll chat like hell."

I dreaded the days when I drove to the Springs for supplies.

Especially I dreaded the visit to the custom slaughterhouse for meat. I learned to avoid going there on the days when pigs were slaughtered, because of their machinelike screaming as they were conveyed along, hanging heads downward. Once I asked the city inspector, watching television in the office, whether I could possibly have seen an incident that in fact I had seen. "Oh, yes," he replied. "That happens sometimes."

The day was a race to get home before dark. I did not stop for lunch. When I reached home, everything needed to be done at once. In the icy bedroom I would change to ranch clothes. I greeted the animals, then tended the five stoves. I shook and scraped down ashes and clinkers in the coal stoves and built up the fires. I was lucky if I did not have to thaw pipes. I fed the animals. The carful of supplies had to be cared for—vegetables stored in plastic bags, cartons of oranges and apples lugged into a room where they would not freeze. Two hundred pounds or more of meat had to be carried into the storeroom, wrapped in packages and stowed in the deep freeze. Then at last I could prepare supper for myself.

Alatna and I were at ease with each other this winter more than we had ever been before. I had thought no further depths to our relationship were possible, no more learning, nothing new that Alatna had to show me. Partly the ease with her and the wolf-dogs was due to the absence of hostile animals from the pen. So partly the happy ease was due to me. For years I had been daily familiar with the feeling of "bristling" on thighs and backs of arms—with guarded fear. Now perhaps it was less that the wolf-dogs had changed than that I had changed. I "pitched it to the wind" lightly with them. I was happier with them. I knew them so much better. I knew they did better with a light touch. My voice pleased them better.

One morning Alatna gave another sample of new behavior she had been showing for the past two weeks. She wanted to get the dogs, especially Toklat, a potential mate, out into her own territory. The wolf-dogs would kill Toklat and the others if she succeeded. But for her part she just wanted to run with Toklat, court and be courted. She was trying for a "fiancé" against her coming estrus.

What she did to effect her purpose was to hold the screen door open with one paw and throat-coax the dogs. Or sometimes if I held the screen barely ajar, instead of coming indoors she backed off, ears "watching," and coaxed the dogs to come outdoors. This they would have done. So there was turmoil at the screen. I barred the surging dogs with my arm. The big wolf-dogs, heads low, eyes bright and ears tipped forward, tried to brave whatever risks they felt in the situation and shoot indoors past me. Meanwhile Alatna was coaxing, there were barks and general excitement. I had a clear vision, aural and ocular, of a dog being slowly murdered. And Alatna, her purpose being frustrated by me, looked up at me black-eyed with will to punish me, and with no subservience at all. When I smiled and looked at her eyes she would turn and jump on some wolf-dog near. Thus I got a daily emotional workout before returning to my cooling coffee and scrambled eggs.

On this morning, after two weeks of attempting daily to coax the dogs out, Alatna first gave me a look recognizing my intransigence about her efforts, then she jumped at my face and kissed it, her eyes radiant. (There is nothing so gracious and sweet as a wolf's kiss, because she kisses with her eyes, her ears, her whole self, as well as with her tongue.) As I soon realized, not once thereafter, not one single time again, did she coax the dogs. She might hold the door open for a minute before entering, but never again whined, never backed out invitingly. With that sunny kiss the wolf had given up. Wolves are great little terminators. Alatna had ended this affair with a flourish.

She had yielded to me on the score of getting Toklat as suitor. But twice during the winter, in her role as matriarch, she controlled me, by methods rarely used. She gave me a review course in them.

On a day of new snow in sunshine I took the long unused

camera outdoors to expose some old film. Shy, beautiful in his winter coat, was coming up from the big pen. I lifted my camera to my face. Uneasily he moved away. He did not like that dark object before my face; he had seen something like it when Noluk died. Nevertheless, I followed him.

Overtaking me, Alatna stood upright beside me and shouldered and fawned with such an energy of radiance, such a force of gaiety, that it broke through my obtuseness: she was trying to control me. I felt controlled. She did not like my following Shy with that dark gunlike object. She was protecting him from me by exactly the same means as she had once used to protect me from him over a dead sheep. (Always she defended the weaker, the threatened one.) I gave up and took the camera indoors.

Later, on a gray evening of damp black earth and melting snow, Musa took from the counter back of the door a package of noodles I had laid out for a dinner of noodles with vegetables and cheese. It was the last package I had. I tried for it. Musa ran down to the big pen. I hurled the nearest object, the aluminum dipper from the water pail, ahead of him and ran after him. Beside the shelter I picked up and tossed bones from the collection the wolf-dogs had made there. I was trying the old "golden apples" technique. Musa ran with the other wolf-dogs to examine each object thrown but he did not lay down his trophy; it was too light, he was too highly elated by the commotion.

To Alatna my action toward him must have looked like aggression. Coming up to me and seating herself on a wide base, she bowed deeply to either side, a striking gesture. She whimpered and begged. I remembered another occasion when I had witnessed such an extravagance of self-humbling. It had been when Cris walked toward her den of newborn puppies, and it had been followed by leaps at his face. Anxiously I knelt and "explained" to her that I meant Musa no harm. Appeased, she went over to the shelter and lay down but continued to follow my actions attentively. As a precaution I turned to her every minute or two and deprecated what I was doing.

Musa won. I chose something else for dinner.

Why did Alatna use the two different means of control toward me? I think she used the radiance control, an outpouring of energy, when she felt full of energy and the humility control when her energy was depleted, as by pregnancy or newborn puppies.

23 *Mated*

WHENEVER the door of Crag cabin opened, it opened onto wild beauty. I stood at the door one morning in late February and just looked around.

Space was not level here. It was deep. It was measured by the spruces, seventy feet high at my right, matched by the ponderosa pines on my left as I looked upward. It was measured by the white cylinders of aspen trunks. Above them by the sunny banks of evergreens and above those by a dark spur of trees in shadow, and by the depth of space still higher, hollowed and held between the far and high skyline cliff-teeth and the spurs and steps of naked crag below.

This third dimension of space was measured constantly by the birds crossing. A raven, white toes tucked to shining body, went behind a theater wing of aspens. Magpies swung across. Something dark blue, a jay, passed almost down on my level, where I stood at the door. The height of space was measured by the bird cries entering it from the right, from the left, from the top, where sometimes a hawk whistled.

Space was a structure here. Invisible ramps descended the air. The birds went up and down on them, but not too high, not clear out of their own layers of air. Down on the ground were the impersonal juncos; the finches huddled together like a crowd of little hens.

All the day of March 13 everything had a quality of pleasure, though yet it seemed average. There were three especially happy moments, each so average on the surface, yet so strangely happy.

The first was at morning. In the webs of bare aspentops, gauzy with new snow and the first sunlight, sat small black-looking forms, scores of gray-crowned rosy finches, every one focused on the birdseed I was showering on feeding trays and the swept terrace in the courtyard. "Here, bird," I called, and half a dozen jays, not collected but scattered in the spruces, watching how intently, could not help cawing with excitement.

Dogs fed, wolf-dogs fed, myself fed too, I went down the snow trail among blue-dimpled wolf tracks to the hay shelter under the spruces with tidbits to distribute purely for love and joy, not as food.

A bite to Musa, lying alone at the front of the shelter. One to Shy. The biggest sliver to Alatna, lying in the deepest, softest hollow of hay. Luxuriously she dropped it, accepting attentions from all the wolf-dogs. Luxuriously she ate it when I presented it again.

She rolled over, her white-furred belly up, forepaws hanging, while Bakno kissed her face and all the others kissed her here and there from tip to tip. Tails wagged; ears were laid sweetly back. Eyes brightened. The suffusion of friendly feeling among the wolf-dogs was almost tangible, was like a field of force. Shadow-eye, whose eyes toward me were usually neutral and dull, gave me as bright a look as to Alatna as he kissed my hand, including me in the general overflow of good feeling. Good feeling was exchanged on all sides. Alatna, the center of it all, the incitement of it all, luxuriated in the wolf-dogs' affection.

"It's worth it," I thought this one more time. Alatna's life, turbulent, dramatic, deprived, had made my own the same. All winter the price had been too hard in terms of human loneliness and work. But a moment like this reassured me.

On this morning the usual screams from Shy that meant Alatna jumping on him were absent. Because the animals were so silent I went several times to the bench to see if they were really there or had escaped. In the middle of the morning I found them lying in sunshine on bare dry ground near snow, under the big fir tree, a former haunt long unused. Alatna's fur was warm with sunshine to my touch. The young animals, all but Musa, lay close to her, and Mr. Shy right beside her. He growled. At me? No, at Musa, who had stepped up behind me.

The best moment was at evening when, again checking up on the silence, I went over to the bench. Already the sun was down here; shadow covered the bench like a pool. The draw I had climbed from was gray in shadow. But up here there was a quality of light that lifted the spirits and made one feel happy. The air was infused with warm gold color. Aspen trunks, rising white into sunshine, reflected light into this reservoir of amber air. Beside them a rock face shone toward the sun I could not see. Above it, higher crags blazed to the level gold. As I walked along in this pool of shadow that was not gray but infused with sunlight, it felt homelike and welcoming, surprising one with ease and pleasure of spirit.

The next morning I stole over to the bench again to see why the quiet, to govern the fear "Have they gone?" until I should see. And Alatna and Shy were standing up "dancing," arms around each other's shoulders. When they sank down after numerous dances, she struck his shoulder with her paw and they rose and danced again.

"It's the way she and Baranof danced," I thought innocently. "And the way she got poor old Taffy to dance with her."

Shortly afterward—one more anxious visit over to the bench, meeting on the way Shadow lying quietly alone—I saw that there would be puppies in the den once more.

Returning to the cabin I counted sixty-three days on the calendar. And this time Alatna would have a wolflike mate who, as with Yukon, would stay with her, give her all the comfort and attention she had craved in previous pregnancies. Already he had helped her dig a new possible den. (More for both to go through the ancient ritual than actually to occupy.) Shy would care absorbedly for the puppies when they came. He would help Alatna feed them and would stay near them, watchful.

When he and Alatna sashayed up onto the plaza soon afterward, Alatna was consummately happy, without alloy. Mr. Shy was beside himself with delight, fending off the others, who no longer dared come near Alatna, except perhaps for a sniff by stealth at her fur. Shy was all king. Pride and brilliance were in his stance and bearing and in his slant-looking eyes. He stood beside Alatna. She again as formerly was mother-queen. He was the chief. They all accepted it now. Musa, Bonnie-face, Shadow, they conceded.

24 *Russian Roulette*

RUSSIAN roulette began the next day. A storm grayed the mountains to receding shadows and terrified the day with wind. Gusts blew up to a hundred miles an hour, according to the radio weather report. The danger was of a tree falling across the fence and breaking it. The trees had grown tall since we had come here.

I did what I could to prepare for the roulette. I kept a warm jacket at hand all the time, whichever building I was in, its pockets stuffed with wool scarf, knitted cap and lined gloves. On the counter by the cabin door stood an electric lantern ready to be snatched. Beside it I laid a sack containing saw, ax, a coil of bailing wire and heavy clippers. If I heard a tree crash I would take these things and run to try to find the spot before the wolf-dogs ventured to it. If the fence was broken I would try to mend it. If I could not, I intended to stand in the breach and fend off the wolf animals from leaving, for as long as I could hold out.

That evening Cris came home and the burden of vigilance partly unrolled onto his heedless cheerful shoulders. That night I slept because there were two to share the alertness. The wind continued the next day. The air was thin and colorless. Rocks and sky looked flat. Cris left in the evening and I lifted the burden of aloneness and responsibility to my own shoulders again.

The next days were alternations of danger and calm. "Wind dangerous and intense" was often the forecast. This was the worst spring in sixty years. There was peace at times; my heart unwound. Then I felt the stir of dull dread before I noted the cause—the rushing of wind again, a Niagara cataract, firs flinging their boughs, everything in commotion. Blind, half-sleepless, I played Russian roulette night and day, not with my life and not with one only of the wolf animals' lives, but with the group of them. If one went, all would go. If all went, none would return.

Alatna's heat was soon over. It seemed short, perhaps because it was so late in her season when she accepted Shy. Now sometimes she rolled him down screeching for a minute. What terrible conflict in both of them. If Shy had not known a mate he would never have had the determination that had overcome his "awe" of Alatna.

Now that her heat was over he slept near her but not touching her. He ate, he was hungry. He, who two days before had been as fierce and dangerous as the storm, came up to the cabin of his own accord one evening to make social contact, rear his head and talk, "*Eh*-oo, *eh*-oo," a genial, emotional sound.

I sat at breakfast the next morning facing the picture window, not precisely alone. On the couch back of me the dog pup Frazier, awakening, confidently mentioned "Ow-r," on a yawn. Big Toklat, on the bunk among the other dogs, scuffled a snore. These things were all but unnoticed, the mere hum in the background. My eyes were unconsciously busy in front, where the feeding trays and swept courtyard terrace were active with life. A jay swooped wide wings without bothering to alight over a flock of finches. They swirled up and twittered. The shy alert nuthatch came alone, and the chickadee.

Above them, above the darkness of firs spiring along the

white brook down in the shadow, rose aspen tops still bare but different-looking this morning, against the pale, gauzy sky. Through their web, beyond it, rose a Tarryall peak, just across from the foot of our draw, grayed with new snow from the night. There was nothing very pretty or beautiful about it all. I thought of friends in Los Angeles looking out from their breakfast room over the stucco-walled rose garden and the descent of wooded hillside toward the bluish plunge of smog, through which faintly would show the far-off white buildings, the great city. But I, looking up at this quiet Tarryall peak, new-dusted with snow, felt, "It's holy." Two finches tossed at each other, a second's fuss. It was not unholy.

What made me calm and ready for delight this morning was the weather forecast: no wind. Those grim days of wind shook me yet. "So, Lois Cassandra," I told myself, "not all trees fall. They're built to stand. I play Russian roulette but I may win."

Before starting the day's work I ran down to the shelter. The wolf-dogs had long been up, but Shy and Alatna had gone back to sleep in the shelter. They lay each curled in a hay hollow. Shy's eyes rolled toward Alatna as he wakened, then aside toward the wolf-dogs gathering happily on his other side. The two chiefs—Shy, Alatna—gratified, accepted their crisp-fried breakfast sausages, then rose, stretched and stepped to kiss lightly each other's muzzles.

What did I do all day? Today was an easy day because Marie came. We worked together at cleaning up. After she left at two o'clock I started the nervous round of dog walks. There were three dogs that would run away if any two of them were teamed. So three rides were required, to a safe run-stretch of road. It took time.

Alatna lay in the yard by the rock wall, an old haunt. In the gray afternoon light her eyes were green orbs, inspecting mine at face-to-face distance. Brown was diffused around her pinhead pupils. So: she wanted attention again, now that her ball was over. She also wanted food and I started broiling in the oven trayful after trayful of evenly sliced liver. Marie had the whole beef liver cut for me. Time. Cool the slices, cut each in three—Alatna choked on whole slices. Even-handed parceling out, piece at a time, first to all the wolf-dogs, then to all the dogs, who, remarkably, lined up shoulder to shoulder, head alongside head and stood still while I went down the line with a bite apiece, then started over. (Liver was too expensive to be turned over for a few to hog.) Time.

Somewhere I got in my lunch or dinner. And always there was the murmur of affection, understanding, rebuke to all as I passed in and out of the yard. Shy irresistibly found himself near time after time to have his neck fur divided and groomed of burs. Also when I went to Alatna, irresistibly myself, he was silently alongside me in a moment: "Pet me."

"Love them now, respond to them now," I thought. It had been so short a time, so long a time that I had known them. "Will you come back," I wondered, touching Shy's mystery-holding head under its soft fine fur, "if the fences break?" I was more tender, more forgiving than I used to be; I appreciated the animals more. Their affection—I valued that more. I used to take it for granted.

Now, shake down, shovel out, carry out ashes. Three coal stoves. Drudgery. Time. But today much of the heavy work I usually did was already done. Cris, before leaving, had carried five buckets of coal to the cabin, the same to the two greenhouse coal stoves. He had filled the oil tank of the stove in the storeroom and set at hand a five-gallon can of oil drawn from the drum on the roadhead.

Still, after the days of roulette, I felt listless. Under the murmur, hardly breathed, barely intonated, of love and of my constant pleasure in all the varied eyes and askings, began to rise the dullness of fatigue. A harsh feeling came over me. Why? The wind. No, it was only the deep-freeze starting up.

In the gray light the horned owl hooted, his first note for the evening. The sound made the pure twilight air occupied and beautiful. There was no answer: the other male had been absent or dead for two weeks now. And for years no female had rippled her extra hoots in answer, so comfortably complete, so satisfying. The male, hoo, hoo-oo, hoo, hoo. The female, hoo, hoo-hoo-hoo, hoo-oo, hoo-oo.

Other wildlife, once common, had suffered change since our coming here, though, except for the magpies, not because of us. The "world" here was wide. The fence divided. Those within were captive. The untrampled ground lay outside, full of fresh enticing smells, all exhilarating and promising to an animal stepping along on it. It was covered with carpets of kinnikinnic, with dead leaves and new herbs. But the wildies that should have varied it were dead—the coyotes, the bear, the cougar, the dainty wildcat. The coyotes had been poisoned or trapped. The others had been shot. Deer, elk and bighorns were diminished or gone, through hunting and poaching or disease. The eagle no longer soared around the skyline crags. Because the magpies had come for the animals' meat, the hermit thrushes had left.

So, infallibly, the two were twined, happiness, sadness and, these days, dread of wind.

I was glad to take books and come up alone to the "hill shack," my retreat on the ridge side above the roadhead. Because of Marie, her young strength, I could escape now to my den, to radio and books and the easy warmth up here of the gas stove.

There came a short quick sound that might mean "escape." It was still twilight, a flashlight not needed. I put on my lined jacket and started the round of the fences.

Feet hurried behind me. A big gray body overtook and hurried alongside me, crowding against my leg—Alatna. I stopped. She seemed as tall as I, as she stood up to kiss my cheek. Her eyes in the gray twilight were green, mottled slightly with light brown. This evening they were not changed or overmantled by any urgency. They were not brilliant or black. One expected transparence but instead they were an inner surface of this varied color, completely themselves, in this light and this mood. Serious, steady, the great greenish mottled orbs looked into mine, inches away. They were completely Alatna, not trustful but something more complete and unshaken than that—my own Alatna, my friend. Shy was right beside her. I crouched and his black nose dotted ice on one cheek while Alatna polished the other. So vulnerable, so confident, all of them.

Going off down the avenue alongside the black trees, my boots crushing the thin icy snow, I glanced back. Shy usually followed me around the fences. He liked the expedition. But now he stood back by the shelter, gazing after me. Returning from my rounds I saw why: Alatna lay there; he was staying with her. The wolf had a "husband" at last, not a touch-and-go mate but one who would stay with her, choose her, like to be with her.

To bed at last. Doonie lay on her canvas beside my pillow. The red curtain rose from the corners of her eyes, slipped back, came up again in irresistible sleepiness over the moonrise of her black pupils. Her white forepaws were curled. They jerked. She slept, with the faintest of deep breaths, not quite snores.

25 *Spring*

Spring was dingy this year. There had been no rain since the previous fall. Even the mountain bluebirds could hardly stand out from the tan of the pastures. In the big pen the woods lacked the green herbage that should have been appearing. The wild lilies of the valley along the brook were not coming up. The brook was low. It dried up before reaching the road crossing at the foot of the draw, where always before one's car had splashed through a pond. On the ridge the lichen clumps lay in their patterns on the kibbles of red granite, but they were ashes; they pinched to powder in one's fingers. In the outside woods the anemones had not lengthened out; the tulip-sized flowers crouched in their vases of leaves.

But there were the little efforts of life. In the outside woods one day there was a bird song, a real song, not the preoccupied noises of busy birds that I heard around home.

The wind still rose at intervals. I hated to leave home for long, even to drive to the Lake for mail, a two-and-a-half-hour trip. Two trees blew down but not across the fences; one of them fell across the road, blocking it. I thought I could saw through it if I had to drive out before Cris's weekend return.

In spite of the frequent turbulence of the weather there was peace in the pens. One still afternoon all four young wolf-dogs and Shy romped at the big open play place at the far end of the avenue. It was a two-ring circus. Musa and Shy played together in the foreground. Shy snapped his teeth at Musa, inviting him. The two wolf-dogs flew up into the air, to land and speed away and back. Shy shook himself so loosely and thoroughly that his lip was flung upward, baring his teeth.

In the background the smaller three played, running, leaping over one another. One of them thought of drinking. At the pool in the woods by the play place, silver lines showed on the dark surface as the wolf-dog lapped. Instantly all the animals were thirsty. They crowded the side of the pool, their tails angled out at "Happy."

In the low yellow sunlight Shy trotted off up the avenue, going to pay attention at Alatna, lying on the plaza. He trotted at a limber jog, his big paws flipping backward in perfect, instantaneous relaxation. His eyes and bearing were purposeful. There was brightness in his eyes.

His return to check up on reposing Alatna was exactly like the action of a wild wolf I had observed on the Yukon once. Nine wolves had trotted upriver to where Cris and I stood, on the bank. A few lay down. A few went on upriver. But one of these had left the party and returned to a reclining wolf and nosed her, a social, meaningful act. Had she been his pregnant mate?

Alatna rose. The wolf-dogs had come part way up the avenue. Now, seeing that she was getting up for the evening, they bounded toward her, running up along the avenue. Bakno hesitated in the gate of the big pen to greet me, but Alatna was waking up, Alatna had to be greeted. Bakno gave me a bright look and ran on toward the main ceremony. As the wolf-dogs reached her, Alatna became the center of a soft happy commotion. As always, her social life was rich.

One April evening the wind stilled after a day of dust from the high plains to the east. Dust dirtied the air, the tan sky, even here in the mountains. The air smelled dusty. Grit was in my eyes. The mountains were dim forms receding into paleness.

Dark twilight came too early by the clock. Suddenly the stillness felt menacing. Perhaps the feeling was due to the evil threatening day of wind, of bad air to breathe and the strangeness of dimmed-out mountains, mere ghosts of mountains. All at once I remembered a night in the Olympic Mountains in Washington state when I had camped alone in a forested canyon, awaiting Cris, and had felt oppressed in this way. That had been the night when over a hundred fires had been started by lightning in the dry forests. The forests around me now were very dry; there might be a thunderstorm. "Fire hazard extreme," said the weather report.

I drove the Scout up to the door of my hill shack for quicker loading. I piled valued belongings into cartons. Swiftly I sorted my books: this I take, this I can do without. The animals? The dogs I could perhaps crowd into the car. The wolf-dogs? I knew they would not take sleeping tablets. They would feel my excitement. I could not hide it perfectly. Moreover, I would not have time to try to give the tablets—not even to count them out or wait for them to take effect. Gates open. That would be all I could do. I got the sleeping tablets from the medicine box though, and laid on the counter in the greenhouse things I would not wish to forget. But how many things would have to be left behind, calling instantly for replacement. I tried to foresee what I would urgently wish I had the next day, comb and toothbrush, for example. I was deathly tired from anxiety when I quit preparations and went to bed, regardless.

Later I learned that on this evening Cris, too, had considered his position in the midst of dry, uninhabited forests.

On the last evening in April came the first spits of rain since the previous fall. A few damp spots appeared on the plank sun deck. A puzzling noise sounded for a minute on the greenhouse roof. Oh, yes, I recalled—rain. It was as if the weather were shifting gears, getting ready for Alatna's giving birth. She had lain heavily in the yard, passive, for days.

All at once something she did made me realize it was time to clear the rocks out of her den. She was upset. There was a screech; she had Shy down. He yielded, lay paws up. It was not easy to be the half-wolf "husband" of a wolf. I thought Alatna would not have been so frustrated and peevish with a pure-wolf male, nor even with a half wolf if he had not been her son.

Her eyes looked distraught. What made me realize that giving birth must be near was her guarding a turkey leg I had given her, and springing out, teeth bared, at passers. It might be a few days yet, but soon she might not let me work on the den. I put on a denim jacket and crawled to the den for the first time in three years.

The tunnel was narrower and lower I remembered. She had worked on the den. The rocks were gone. The ground was a dust heap. She had dug hard to excavate the old obstacle rock between the middle and back chambers. I was able to move it out through the "window" of the middle chamber. With gloved hand I shoved out loose earth. But I was nervous. If she got the gate of the home pen open, as she could easily do, and was upset with me for what I was doing, I was in a poor position.

A noise sounded too close back of me, as if an animal had got into the pen. It took quite a bit of wiggling and maneuvering for me to back out. But the gate was still fast. However, Alatna had come to it. I was relieved when, as I opened it, she galloped past me, followed by the other animals, to see what I had done to her den.

One day in early May Shy acted more confident than ever before toward Alatna. He stepped over to touch his nose to her face as she reclined. He performed a quaint pure-wolf gesture; he laid a forepaw on her forehead and held it there. Then he stepped a hind leg across her and stood over her for a minute as she still lay there. I had never seen him take this confident position before. Kotzebue had done it from the time he was a pup. Kotzebue had known she favored him. Now Shy, too, was sure of her favor. He was securely her mate-companion, the first she had ever known.

One morning Alatna performed yet one more new action, after all these years. It was slight but impressive. When I went to the pen she came up to me, not talking or especially glad, but *smiling*. I have often said she "smiled." But now for the first time she smiled as a dog may do: the corners of her mouth were tucked tight so that little ripples appeared in the skin, as in a human smile. With that curly tightening of her lip corners, her total ease with me—perhaps with that came the climax. It came not on the live eager little fur balls soon to be in the den with her. It was something unforeseen. All these years it had not occurred to me that she could do this.

The smile surely connoted inner changes. I had not dreamed until now that I had had something to give Alatna—and she had been able to receive it—besides my main effort and preoccupation, which had been, so passionately, to keep her as herself, to keep her heart confident and free. But I had given her something—I did not know what, in a wolf's mind. I had given her of my humanness. She had given me of her wolfness. We were both different. She was still all wolf. I was, I thought, more human.

This change between us was as subtle as light. Perhaps only a spectator could have told what it was, if it had to be described. But only Cris glimpsed us together and even he did not see the yielding, the affection when we were totally alone, Alatna and I. What he saw was the confidence, the dependence. It was a wonderful combination she had—her authority toward her own kind, yet her emotional need of me.

I had "paid" for it. I wished yet this one more time that I could understand this business of "paying." All these years—ten now that I had lived with a wolf or wolves—I had been aware of it. I had felt that whether it was explained or not, it was one of the most important things in the world, to "pay." I had watched in my reading for glimpses of other people's understanding; someone must fathom "paying." And all at once here was a measure of understanding: to pay is to live, not recoil into words. Not to pay is to be detached. "Pay" means involvement, concern, action.

On a gray still morning when the air was warm, the ground dusty, the country tan and dingy with thirst, I looked out the window at all the spars of white aspen trunks in the big pen, slim, together, virginal, holding up tops that were tinged the barest of greenish colors, an almost imperceptible greenishness among the firs, the crags. No, they were not colored. The green was an illusion. Why, yes. It was there!

Impulsively, happily, I left my work and drove to Cris's mine. But what Cris and I talked of was sad. There were personal differences that could not be resolved. Crag cabin would have to be given up, and was to be sold. This meant the end of a home for Alatna. We discussed her fate. No good new pen could ever be built. The building of the pens at Crag cabin had been a once-in-a-lifetime labor. A meager pen, on hot dry stony ground, would leave Alatna crying and disconsolate. There would never be another place like Crag cabin, with its big wolfproof pens holding a running brook, woods, high rocks to climb and a ground cover of wild herbage.

It was unthinkable to turn Alatna over to a zoo, after the relatively free, abundant, social life she had known. She would be hopelessly unhappy. She needed her family; she needed space to run in, variety and the choice of seclusion if she wished it. The only kindness I could give her now would be a quick out—quick unfearing death before she knew misery and privation. It was my obligation to end her life while it was still good. Not only she but the other animals as well must go. There would be no home for them either. They could no longer be sheltered and cared for.

"The Medea decision you face with the animals is awesome," so Carol, our friend, had written. "How many years have you spent finding solace and joy in them, punctuated by periods of such anguish over them as I can only begin to fathom?"

Alatna was nine years old now. During all of those years I had lived with her, except for the first weeks, when she was a wild puppy in a wild den. Seven years and seven months had passed since my "pact" had given to her seven years of my life.

She had never been a pet. A wolf is not a pet. Given the conditions, a wolf can become the friend of a human. The conditions are three: space, four-footed companions and civil treatment. Alatna had received all three and she had become my wolf friend.

We had not willingly chosen captivity for her. We had done so because the choice had been captivity or death. Had we done right to choose captive life? For Alatna the answer was surely yes. Her life at Crag cabin had been as good as a captive's life could be. She had been vital and authoritative, and for the most part happy. At times she had been actually free. As I had longed to do, I had preserved this one animal's sense of autonomy. It was an honorable achievement, though small in a world where wild animals everywhere are being pressed toward loss of habitat and extinction. It was a testimony to the importance of autonomy in a big intelligent animal. All creatures are entitled to their autonomy, within the bounds of the welfare of the whole.

Manipulate or love. We had manipulated Alatna to the very least extent possible, after the big, inevitable limitation of the fence itself. This much had been achieved: Alatna had kept her "pride."

As for the other animals, all were tied together. They needed one another. Toklat was needed by Shy. Shy needed an opposite number, to exercise his maleness, his authority. He came to the fence whining, his head high, his eyes bright and longing, wanting Toklat to come and be challenged. Tommy and Frazier, the dog pups, needed each other to play with. Doonie was a "wife-dog." She and Toklat were affectionate all the year around. Often she went up to him as he lay and, with the other dogs joining in a semicircle around her, began the "Let's love

Toklat" ceremony. She would crinkle her lips back, baring her teeth in a complete dog smile, meanwhile wagging her tail and rump, and barking. Toklat had lain beside her as she gave birth and had helped lick the cauls off the puppies. All of the animals were interdependent. Homeless now, all must go.

I drove homeward sorrowfully from the mine. The next day at Colorado Springs I bought meat for the animals as if they were to live forever. But also I bought sleeping tablets, prescribed by the vet "for euthanizing animals." I was in a state of conflict.

After two days I held out sleeping tablets in the hand that had often held out love.

Alatna went under quickly. Mr. Shy did not. Until he did, the coup de grâce could not be given to all the others, a bullet to the center of the furry foreheads. There was a pause. Alatna lay unconscious at the side of the shady avenue. Slowly the sunlight began to move across her body.

I thought of her free with her fellow wolves on the tundra in the old big days of her youth. I thought of the travel, the speed, the riffling fur of the young, full-furred wolves. Confidence marked their bearing. (With what pains I had preserved that confidence in Alatna.) The wolves were tundra masters. Authority was one component of their cells. Even sitting carelessly, Alatna bore herself with authority. If the grizzly has delight in his land he hardly shows it; pleasure shows mostly in his minutes of relaxation. But in bearing and movements the wolves revealed delight. Their eyes expected it. With swinging exuberance the striding legs laid down each big forepaw as far forward as the nose-tip or beyond it.

And always the tundra surrounded them, full of things to interest them. The freshly snowed mountain walls, that first autumn of Alatna's life, went upward white from the flowing tan tundra, not far off but looking strangely near at hand now that the snow had come. Always the wolves' eyes sought one another; the ears "smiled" backward or "looked" alertly. The intermingling of social responses was a constant symphony played among themselves, with equal sonorities sounding in their nerves from the light, from the delicate fine air and from the pressure, changes and scents of the tundra cover itself.

I was included in the friendliness. The wolves had no intention of starting their daily walk until all had clustered around me, each one to be "loved." Ears sloped back, paws were lifted for good will and not locomotion. Once Mr. Arctic innocently raked my cheek with his claws, standing on hind feet to be hugged. On the walk two or three would rush up to kiss me. There were the friendly little nose bumps.

The wolves examined and explored. When the first thin marginal ice appeared on the tundra ponds, the first ice of their lives that they could walk on, they investigated it attentively, stepping on it and breaking through, looking at it, pushing it with their forepaws. One wolf broke off a piece and carried it eight or ten feet before dropping it.

Big as they were, the young wolves had still been dependent on home. Once when I was far out on the tundra with them a snowstorm came in fast from the north, blank and white, concealing the mountains. The wolves left me and streaked for home, our camp.

Now Alatna's lifelong home was failing her.

For Shy death, when it came, was instant. Then at last I could go to the body of Alatna, panting in the sun, and give her, too, death.

About the Author

Lois Crisler was born in Spokane, Washington, and educated in the schools of her native state. She was an instructor at the University of Washington before her experiences with wilderness life. Her adventures with the wolves of Alaska were the source of her first book, *Arctic Wild*, which was written in part on a grant from the Eugene F. Saxton Memorial Trust. Mrs. Crisler completed the writing of *Captive Wild* on a Guggenheim Fellowship.

About the Author

[illegible]